MW00845652

TAPE CASTING

THEORY AND PRACTICE

TAPE CASTING

THEORY AND PRACTICE

Richard E. Mistler and **Eric R. Twiname**

Published by The American Ceramic Society, 735 Ceramic Place, Westerville, OH 43081

The American Ceramic Society
735 Ceramic Place
Westerville, Ohio 43081

Printed in the United States of America.

04 03 02 01 00 5 4 3 2 1

ISBN: 1-57498-029-7

Library of Congress Cataloging-in-Publication Data
A CIP record for this book is available from the Library of Congress.

For more information on ordering titles published by The American Ceramic Society or to request a publications catalog, please call (614) 794-5890 or visit our online bookstore at <www.ceramics.org>.

Contents

Preface

Fifty years ago, Glenn Howatt first described the process known today as tape casting.[1] Although Howatt used this process to manufacture millions of capacitors, the authors are convinced that even in his wildest dreams, he did not foresee the impact of this technology on the materials processing industry of today. Howatt's invention was born out of the urgent need to find a substitute for mica as a capacitor material during World War II. Tape casting has matured into a full-blown manufacturing technique for the formation of thin, essentially two-dimensional sheets of just about any material that can be produced as a powder. In today's arena, this includes products ranging from metallic electrodes for molten carbonate fuel cells to organic lithium polymer battery components and the more common standard ceramic substrate materials. The technology is even being used to manufacture three-dimensional objects in a process known as LOM (laminated object manufacturing), in which thin sheets are cut to different precision shapes by a laser and laminated in a stack to fabricate a part. It appears that the only limit to the utility of tape casting is the imagination of the materials engineer.

This book is a compilation of ideas based upon experiment and experience. Some portions are pure conjecture based upon our combined 35 years of hands-on experience with tape castings. In many ways, it is a how-to book that details our observations, interpretations, and solutions to specific problems. We hope that technicians, engineers, engineering managers, professors, and students will all benefit from reading this work.

Acknowledgments

I would like to acknowledge the many contributions of the engineers, scientists, and technicians with whom I have been associated during my 40 years of professional life. In particular I would like to cite the personnel in the ceramics processing group at the Western Electric Engineering Research Center, where I was first exposed to the mysteries of tape casting. My coauthor, Eric Twiname, deserves an accolade for keeping me honest during our early-morning rap sessions, where many of the ideas put forth in this book were conceived. He also convinced me that there was a book to be written based upon this wealth of ideas. A very special acknowledgment goes to Dr. Harold Stetson, my mentor and good friend, who introduced me to tape casting as a ceramics processing technique more than 30 years ago. He took a very green novice scientist under his wing and taught him the ropes about this process, which eventually led to this book.

I would be remiss if I failed to acknowledge my wife and life travel companion, Elizabeth Brendel Mistler. Betty has wholeheartedly supported me in this and in all of my pursuits and endeavors during the 40 years we have been married. For this I will be forever grateful.

RICHARD E. MISTLER
YARDLEY, PENNSYLVANIA

I would like to acknowledge Jesus Christ, my savior, without whom naught would be; Dr. Daniel Rase, for making the manufacture of a china coffee cup a topic of fascination one rainy summer afternoon—it is not likely that the field of ceramic engineering would have crossed my mind without that lesson. Also Richard Mistler, for a loose hand in management, a patience to be envied, and the heart of a true friend; and Harold Stetson for motivation and encouragement.

ERIC R. TWINAME
STATE COLLEGE, PENNSYLVANIA

CHAPTER 1

Introduction and History

Since the inception of the modern ceramics era, usually identified as during or just after World War II, there have been several advances in ceramic processing technology. One of the newest and most prolific of these advances has been the development and implementation of tape casting as a manufacturing process for the production of thin sheets of ceramic materials. Glenn Howatt is universally regarded as the "father" of tape casting since he had the first publications and patent describing this process.[1]

Tape casting is also known as *doctor blading* and *knife coating,* and under these names the process is well known in many industries, including paper, plastic, and paint manufacturing. The "doctor" is a scraping blade for the removal of excess substance from a moving surface being coated. The technique has long been used in the paint industry to test the covering power of paint formulations. Films of paint a few mils (< 50 microns) thick are uniformly coated on a standard black-and-white background, and the degree to which the background is hidden is measured optically. Howatt's patent[2] was the first documented use of this technique to form ceramics. His patent was for "forming ceramic materials into flat plates, especially useful in the electric and radio fields." This is still the principal application today, although it extends far beyond the uses envisioned in 1952.

The chief advantage of the tape casting process is that it is the best way to form large-area, thin, flat ceramic or metallic parts. These are virtually impossible to press and most difficult if not impossible to extrude. The difficulties are compounded in dry pressing, when the plate is to be pierced with numerous holes because of the increased problem of uniform die fill. Punching holes and slots of various sizes and shapes into unfired tape is relatively easy and essential to the multilayered ceramic packages being designed and manufactured today. The thin ceramic sheets are essentially two-dimensional structures, since they are large in the x and y directions and very thin in the z dimension. In today's technology, very thin is defined in microns, with tapes as thin as 5 microns being reported by equipment manufacturers.[3] The authors have cast tapes on a standard tape casting machine that range from 12 microns on the thin side to over 3 mm on the thick side. The prime dried thickness range for tape casting is generally accepted as being from 0.025 mm to 1.27 mm.

In the ceramics industry, tape casting is most analogous to traditional slip casting. The processes are similar because both use a fluid suspension of ceramic or metallic particles as the starting point. However, there are subtle differences. Tape casting is usually based upon a nonaqueous solvent as the liquid system. In recent years there have been some excellent successes with aqueous-based tape casting systems, but in most tape casting processes, nonaqueous solvents are required, since the drying process is evaporative from the surface rather than absorptive into a plaster of Paris mold. It is interesting to note that Howatt[2] taught the use of porous plaster batts as the tape casting surface, and water was one of the liquid media he used.

The use of a porous casting surface has been replaced in modern tape technology by the use of a nonabsorptive carrier. In the 1950s, the American Lava Corporation developed and eventually patented this advance in the technology.[4] John L. Park, Jr. described the use of a moving polymer carrier as a casting surface in this seminal patent. This was the turning point in tape casting, since it was the first time that the process was demonstrated to be continuous, and dried, unfired tapes could be rolled up on the polymer carrier for use in downstream processing. It also opened up the possibility of roll-to-roll or continuous in-line processing. Since the Park patent was issued, there have been many refinements with respect to slurry formulation and equipment design, but the basic process has remained very close to that original concept.

Figure 1.1 is a schematic representation of a time line for significant developments that have occurred since the first description of tape casting in 1947.

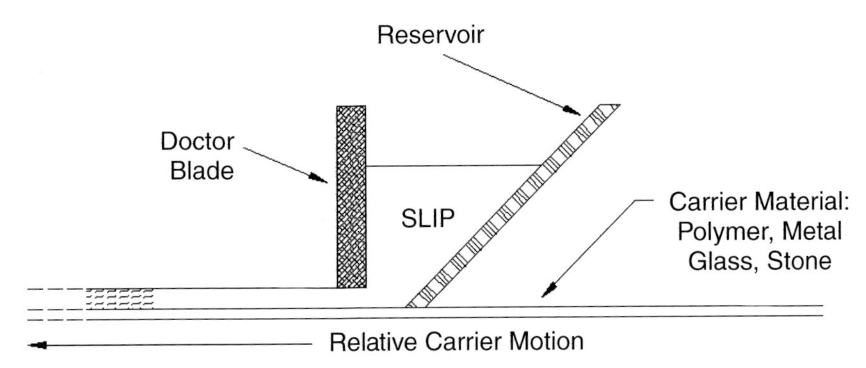

Fig. 1.2 The basic principle of the tape casting process as it exists in most modern factories.

Figure 1.2 illustrates the basic principle of the tape casting process as it exists in most modern factories. It is based upon the principles outlined in the Park patent: a stationary doctor blade, a moving carrier, and a drying zone of some sort. The heart of the process appears very simple, and it really is, but the design of the doctor blades (or rolls) and casting machines, and the formulations that make the process work, are not quite as simple as they appear. Thirty years of experience with the tape casting process have convinced the authors that science alone does not answer all of the questions, but current research is leading us in that direction.

In a typical tape casting process, the slip or slurry is poured into a puddle or reservoir behind the doctor blade, and the carrier to be cast upon is set in motion. The doctor blade gap between the blade and the carrier defines the wet thickness of the tape being cast. Other variables that come into play include reservoir depth, speed of carrier movement, viscosity of the slip, and shape of the doctor blade (to name just a few). The wet film of slip passes into a drying chamber of some sort, and the solvents are evaporated from the surface, leaving a dry tape on the carrier surface.

The remainder of this book will describe in detail the materials selection process, materials processing and characterization prior to tape casting, the tape casting process and equipment, further tape

Figure 1.1

Historical Time Line for the Development and Utilization of Tape Casting Technology

1943–45	First work and first tape casting machine designed and built at Fort Monmouth Signal Laboratory, Fort Monmouth, New Jersey.
1945	First reported results—Report No. 540, Div. 14, N.D.R.C. at Massachusetts Institute of Technology, October 1945, G. N. Howatt, R. G. Breckenridge, and J. M. Brownlow.
1947	First publication—G. N. Howatt, R. G. Breckenridge, and J. M. Brownlow, "Fabrication of Thin Ceramic Sheets for Capacitors," *J. Am. Ceram. Soc.,* **30** [8] 237–42 (1947).
1947	First company established to produce tape-cast capacitors, Glenco Corp., Metuchen, New Jersey, G. N. Howatt, founder and owner.
1952	First patent issued: G. N. Howatt, "Method of Producing High-Dielectric High-Insulation Ceramic Plates," U.S. Patent 2,582,993.
1952–54	Development work at American Lava Corp., Chattanooga, Tennessee, on continuous manufacture of ceramics by tape casting.
1954	Patent filed by John L. Park, Jr. of American Lava Corp., "Manufacture of Ceramics." Teaches continuous casting on impervious polymer carrier.
1958–59	Development work at RCA on multilayered ceramics using tape-cast material.
1960	Patent filed by W. J. Gyurk of RCA on the lamination of tape-cast ceramics to form multilayers.
1961	Patent issued to J. L. Park, Jr. of American Lava, "Manufacture of Ceramics," U.S. Patent 2,966,719. Teaches continuous tape casting.
1961	First presentation on multilayered ceramic packages prepared from tape-cast materials: H. Stetson and B. Schwartz, "Laminates, New Approach to Ceramic Metal Manufacture: Part 1: Basic Process." *Am. Ceram. Soc. Bull.,* **40** [9] 584 (1961).
1965	First patent on co-sintering of metals and ceramics: H. W. Stetson of RCA, "Method of Making Multilayer Circuits," U.S. Patent 3,189,978. Teaches the use of co-sintering metallized and laminated tapes into a monolithic structure.

1965	Patent issued to W. J. Gyurk of RCA, "Methods for Manufacturing Multilayered Monolithic Ceramic Bodies," U.S. Patent 3,192,086. Teaches the lamination procedure.
1967	First presentation on the development of thin-film aluminum oxide substrates prepared by tape casting: H. W. Stetson and W. J. Gyurk, "Development of Two Micro-inch (CLA) As-Fired Alumina Substrates," *Am. Ceram. Soc. Bull.*, **46** [4] 387 (1967).
1967	IBM reports on work on multilayered ceramic packages prepared from tape-cast ceramics for computer applications. B. Schwartz and D. L. Wilcox, "Laminated Ceramics," Proceedings of the Electronic Components Conference 1967, pp. 17–26.
1972	Patent issued to H. W. Stetson and W. J. Gyurk of Western Electric Co., "Alumina Substrates," U.S. Patent 3,698,923. Teaches the use of fish oil as a dispersant for tape processing of very fine alumina powders.
1978	First definitive publication on the tape casting process: R. E. Mistler, D. J. Shanefield, and R. B. Runk, "Tape Casting of Ceramics," in *Ceramic Fabrication Before Firing;* pp. 411-48. Edited by G. Y. Onoda and L. L. Hench. John Wiley & Sons, New York.
1980s–1990s	Many publications on materials development and process improvement. Basically a period in which the technology matured. New applications were being explored, such as the production of thin membranes for fuel cells and armor plating shields.
1992	First announcement of the use of tape casting to build three-dimensional objects was made by Lone Peak Engineering, West Valley City, Utah: the LOM process (laminated object manufacturing).
1992–1996	Introduction of microgravure and slot die coating heads on tape casting equipment in Japan to produce precision thickness in thin tape-cast materials.
1996	First 5-micron-thick tapes for capacitor applications achieved in laboratory in Japan.
1997	Tape casting equipment capable of manufacturing 5-micron tapes continuously for use in multilayered ceramic capacitors introduced to the marketplace in Japan and the U.S.

processing before sintering, applications of tape technology, and an important section on problem solving for situations that the authors have experienced in slip formulation, casting, and tape processing. There is also a glossary at the end of the book that defines the terminology used throughout. Some of our thoughts and definitions do not fit with those that are commonly accepted in the ceramic processing community.

CHAPTER 2

Materials Technology and Selection

POWDER

What is most important?

Rich: Surface area is most important by far. Moisture content would be second.

Eric: No, I think surface area and density are equally important.

Rich: Surface area in m^2/g takes ρ and D_{50} into account—even particle shape.

Eric: Yes, but not enough. There is no single factor to determine a batch. Surface area is very important, but density is just as important. Try whipping up similar batches for $3m^2/g$ Al_2O_3 and $3m^2/g$ WC !

Rich: I think you're wrong, Eric. If you had $3m^2/g$ you could batch them the same way...I don't care if it's WC or SiO_2. I still say surface area is most important.

Eric: Surface area, ρ or D_{50} ... you don't need all three, but you can't get by with only one. You need surface area and one of the other two, preferably ρ because it's easier to think in weight rather than volume.

I guess we disagree. Let's explore the topic.

2.1 POWDERS

In any materials fabrication process, the most important ingredient in batch formulation is the ceramic, metallic, or composite powder. After binder removal and final consolidation, the powder(s) are the only portion of the batch left, and they define the properties of the part produced. The other ingredients in the batch formulation, such as the solvent(s), plasticizers, binder, and surfactants, are there simply to facilitate the fabrication of the desired shape and the green (unfired) bulk density of the formed part. Essentially, the tape casting process is used to obtain and hold the powder particles in the desired configuration so that, after sintering, the final part has the desired size, shape, and properties.

In many cases the powder is the one ingredient over which the materials engineer or scientist has the least control, since it is usually selected to provide specific properties in the final product. In other cases, it is selected for its ability to be processed or sintered under specified conditions. In our laboratory, about 80% of the materials we are asked to process into tape-cast products are preselected by our clients. In the rare cases where we have the liberty to select the starting powder(s), we can define several powder properties to make the tape casting formulation more "forgiving" and easier to process. These powder characteristics will be reviewed in the sections that follow. It is often, but not always, necessary to combine different powder chemistries in order to achieve the desired strength, resistivity, dielectric constant or dielectric strength, chemical resistance, porosity, or other fired property or characteristic.

In any materials process, it is essential that the starting powders be well characterized. This is especially true for tape casting. The important parameters to monitor in all powder lots are the average particle size and distribution, surface area, and trace-impurity level. The powder density in weight per unit volume is also an essential property that must be considered when formulating a batch for tape casting.

2.1.1 Particle Size, Distribution, and Shape

The generalized topics of particle size, particle size distribution, and particle shape have been covered in numerous books[1,2] and articles.[3,4,5] In addition, particle size effects in ceramic processing have also been covered by several authors.[6,7] In this section we will discuss the aspects of particle size, distribution, and shape as they relate directly to the tape casting process.

The green bulk density generated during the tape casting process is unique in materials processing. It is the only fabrication process in which gravity and the shrinkage of the organic system during drying generate the packed density. Even though it is similar to slip casting in that it is a forming process from a liquid dispersion of particles, the capillary force generated by the porous plaster of Paris mold is not present in tape casting, since it is an evaporative process. This unique densification process demands control of the particle size and particle size distribution in order to obtain the highest possible green bulk density. There are many examples of green bulk densities from tape-cast products that meet and in some cases exceed those generated by pressure-forming techniques such as tablet or dry pressing at pressures up to 138 MPa (20,000 p.s.i.). An example in our laboratory is that of a 100% aluminum oxide tape that has been cast at green bulk densities of 2.78 to 2.81 g/cm^3. This is equivalent to greater than 70% of the theoretical density of pure alumina (3.986 g/cm^3). We feel that the tape casting process, if done correctly, can provide some of the highest densities before sintering of any materials-forming process.

An excellent discussion of particle packing has been written by Shanefield.[8] In that publication it is shown that in commercial powders with a typical particle size distribution, the best lubricated and pressed packing factor attainable is roughly 55% of theoretical. The reason for this is that most commercial powders are processed (ground) or precipitated to yield very high percentages of fine particles that do not pack very well during pressing. The finer the powder, the higher the surface energy, and therefore the greater the driving force for the sintering process that must be performed on most formed parts to yield a dense final product. These are typically the kinds of powders that we must use in our tape casting formulations. Most of these powders have a surface area of 5 to 10 m^2/g or higher and a relatively narrow particle size distribution.

In the tape process, one of the most important steps is the dispersion milling procedure. This will be described in more detail later in this book. One of the primary purposes of this procedure is to break down soft agglomerates that have formed as a result of high surface area. Dry agglomerates are caused by weak interparticle forces, that is, van der Waals and H-bonds. If one looks at a scanning electron micrograph of a typical powder in the as-received condition, it will look like the one shown in Figure 2.1. The agglomerates look like fuzzy balls of material with very little surface definition. In order to get an accurate particle size distribution and particle shape, these agglomerated powders have to be broken down to their ultimate size, as they are during the dispersion milling procedure.

One way to do this is to go through the dispersion milling procedure and then analyze the particles in the dispersed state. Another way to accomplish the same end is to use an ultrasonic dispersion technique such as the one described by Shanefield[9] and then determine the deagglomerated particle size distribution. As described by Shanefield, the particle size thus determined is actually an aggregate size, since some of the "hard" agglomerates will not be broken down. "Hard" agglomerates are groups of particles that stick together (or fuse together) to such an extent that they are not broken down by the ultrasonic or the milling procedure. The ultimate particle size and size distribution can be determined by any of a variety of techniques, as long as the liquid dispersion does not go through a drying step. One commonly used procedure is the sedimentation

Fig. 2.1 Typical powder in the as-received condition, as viewed through an electron microscope. From *Ceramic Processing Before Firing,* eds. Onoda and Hench. © 1978 John Wiley & Sons, Inc. Reprinted with permission.

technique, in which the dispersed powder particles settle under the influence of gravity and a scanning instrument such as an X-ray beam or laser detects and graphs the size data automatically. There are many excellent references concerning particle size measurement,[10] so no further discussion will be presented here. It suffices to say that the analysis should be made on a liquid dispersion of the powder to be measured without ever going through a drying step.

In a well-dispersed particle suspension where the "soft" agglomerates have been completely broken down, one can spray a small amount of the slip onto a collodion-coated glass slide and carbon shadow the surface at a low angle ($\sim 23°$) for use in a transmission electron microscope. By this technique the ultimate size and shape of the powder particles can be detected. An example is shown in Figure 2.2.[11] These are actual particles of the agglomerated powders previously shown in Figure 2.1. The range of sizes observed by this technique closely matched the particle size distribution on a dispersion of the powder analyzed by a Stokes' law sedimentation technique using the transmittance of an X-ray beam as a function of settling.[12]

During the evaporative drying of a cast tape, the well-dispersed particles pack together to form a dense bed. It is important that a distribution of sizes be used so that the closest (most dense) packing can be achieved. Small particles fit into the interstices of the larger particles, and the void space is partially filled. In real life, the particles are irregularly shaped, and bridges can form that limit the settled packing density that can be achieved (especially in the absence of applied pressure). Typically, the range of sizes in a powder is not very wide; for example, in super-ground alumina the D_{90} is in the range 0.6 to 1.1 microns, and the D_{10} is in the range < 0.1 to 0.2 micron. The range of sizes for these powders is of the order of $\sim 6{:}1$ for the largest to smallest particles. As shown by Shanefield[13] using a distribution of spherical particles with a range of sizes of $\sim 25{:}1$, a packed density of $\sim 95\%$ can be achieved. Actual particles with a wide range of sizes such as this could be batched, but severe effects on the sinterability and grain growth during sintering would occur. For example, it is well documented that "seeding" large particles in a fine-particle matrix contributes to secondary grain growth in ceramics.[14]

Learning to work with the particle size, particle size distribution, and particle shape of commercially available powders is a part of the tape formulation technology that we all must face. Add to this the wide range of density values encountered—for example, WC with a density of 15.6 g/cc and SiO_2 with a density of 2.16 g/cc—and the problem is compounded. Using the proper dispersant and the proper dispersion milling procedures as outlined in this book can usually overcome the handicap of having to use a powder without the optimum morphological characteristics for processing.

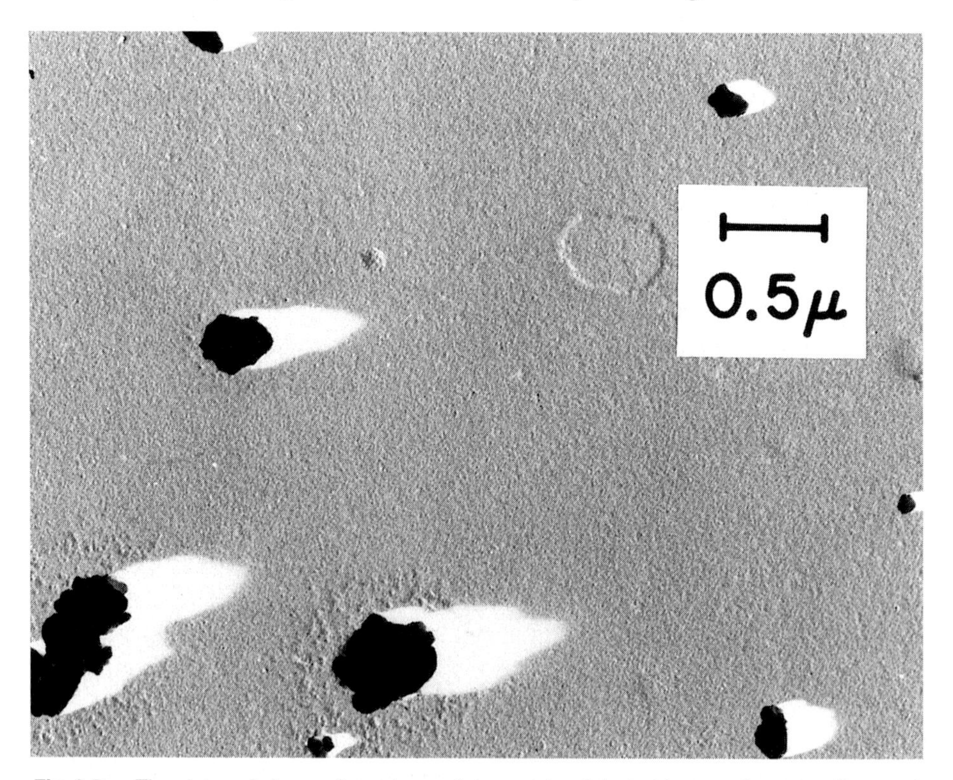

Fig. 2.2 The size and shape of powder particles can be detected by spraying a small amount of a well-dispersed particle suspension on a collodion-coated glass slide and carbon shadowing the surface at a low angle (~ 23°). From *Ceramic Processing Before Firing,* eds. Onoda and Hench. © 1978 John Wiley & Sons, Inc. Reprinted with permission.

2.1.2 Surface Area

Usually the first powder characteristic that we are concerned about in our laboratory is specific surface area (more commonly referred to simply as surface area.) The surface area of a powder is a measure of its size, shape, and irregularity (such as the presence of voids that are open to the surface). There are several excellent books about surface area and its measurement.[15,16] The most commonly used technique for determining powder surface area is the BET method using the adsorption of a monolayer of a gas such as nitrogen on the powder surface.[17] The common unit of measure for surface area is m^2/g (area/unit mass). Most powders fall in the range of 1 to 50 m^2/g, with the vast majority of highly sinterable powders falling between 5 and 15 m^2/g.

From the processing standpoint, the materials scientist should be interested in the surface area of a powder because surface area is the most significant factor in its interaction with organic additives such as surfactants and binders. Processing chemicals are usually added as weight per weight of powder and are best associated with the area they have to interact with to provide the proper dispersion, lubrication, or binding qualities. In most cases, the total powder area is directly proportional to the amount of organics that have to be added to make a process work. This is particularly true for tape casting. From a point further downstream in the processing sequence, surface area can significantly affect the sintering temperature versus fired density relationship.[18]

One can begin a formulation based upon past experience with a powder of similar surface area, but the results may depend more upon trial and error, since factors such as surface chemistry also play an important part in organics selection. In general, high surface area powders such as those above $\sim$ 20 m^2/g are much more difficult to work with than are those in the 5 to 15 m^2/g range. The selection of the proper dispersant/solvent combination, as well as the proper dispersant and dispersant concentration, are all critical factors for the very-high-surface-area powders. These topics will be covered in the sections that follow.

2.1.3 Powder Density

Particle weight is another property to consider in the formulation of slips for the tape casting process. High-density powders, such as WC with a density of 15.6 g/cc, may require the use of a powder with a D_{50} that is much lower than a powder with a density in the 2 to 3 g/cc range. This is required to keep the very heavy particles in suspension during the casting process. At times, powders with very different densities have to be batched in the same formulation, and this also requires tailoring the organics to keep the various components in suspension at the same time. Examples include materials such as alumina, clay, and talc, alumina and zirconia, and zirconia and yttria.

Powder chemistry and its effect on fired properties are beyond the scope of this book but will be mentioned throughout in specific examples. Powder chemistry, and surface chemistry in particular, however, has a very large effect on tape casting slips. As will be pointed out throughout the book, the effects on the slip from one component cannot be successfully separated from the effects of other components. Each slip component will, however, be handled separately to help us gain a better understanding of the individual actions before addressing the interactions between ingredients.

SOLVENTS: You can't live without them ... but they might kill you.

Why is it called a *homogenizer*? Its main purpose is to prevent skinning.

The literature refers to a homogenizer, but what does it homogenize? There are such things as homogenizers, but MEK and ethanol are fully miscible, even azeotropic in some ratios; why do we need a homogenizer? I guess, in a very philosophical way, it homogenizes the solvent concentration throughout the tape during drying by working to avoid the solvent-depleted "skin" on top.

I read somewhere that an expert is someone who not only knows all of the available information on a subject, but also knows which sections of it are wrong.

Enough of the semantics, let's explore the topic.

2.2 SOLVENTS

As mentioned in previous sections, tape casting is a "fluid forming process." The mechanical metering of the powder into a two-dimensional sheet requires that the powder behave (flow) as a fluid. To achieve this degree of formability, we suspend the powder in liquid. That liquid is referred to as the "solvent," the "vehicle," or the "base." The liquid is also used to distribute the other ingredients homogeneously (we hope) throughout the slip to create a uniform mixture.[19] Other ingredients to be distributed may include binder, plasticizer, lubricant, release agent, flattening agent, thixotrope, wetting agent, dispersant, deflocculant, ceramic precursors, catalysts, and the like. While each of the additives will be addressed individually or in groups in following sections, this section will look into a variety of solvents and their effects on the tape casting process.

We mentioned in the Introduction that the majority of tape casting slips are formulated with nonaqueous solvents. The organic solvent vehicle is so widely used, in fact, that it is simply referred to these days as "solvent" or "solvent-based casting," as opposed to aqueous or "water-based casting." Over the years, some organic solvents have become the standard solvents of choice for tape casting slips. These solvents, such as ethanol, methanol, toluene, methyl ethyl ketone, xylenes, and 1,1,1 trichloroethylene, have also become stages for some rather heated debates between manufacturers and EPA or OSHA representatives. We will try to remain impartial in the areas of health, safety, and environmental impact when discussing these solvents and suggest that, in an area as subjective as organic solvents has become, everyone should familiarize themselves with any chemicals with which they come in contact.

The vehicle is called a solvent because it must dissolve the other slip ingredients. The selection of a solvent is, to a large degree, limited by the other additives chosen. When you design a slip recipe, you choose the powder first (see section 2.1), followed by the binder(s). The next ingredient to be specified is a solvent that can dissolve the binder(s) chosen. The solvent must also dissolve the surfactant(s) and possibly some other ingredients. In order to dissolve the wide variety of ingredients needed for other functions, it is common to use more than one solvent as the fluid vehicle.[20]

There are many advantages to the use of multiple solvents in a tape casting slip, and thus dual or "binary" solvent systems are very common in organic solvent–based tape casting. Most of the published formulations use binary blends of organic solvents although we have seen as few as one and as many as four different solvents used in a single slip recipe.[21] The main advantage in using multiple solvents is the increased ability to dissolve.[22] Other important advantages gained by the use of multiple solvents include greater control over drying speed, rheological control, cost, and safety. Many people over the years have also been quite taken by the idea of azeotropes, which will be covered later in this section.[23,24,25,26,27]

Looking at the solvent picture from a casting perspective, the requirements for the solvent(s) are to: dissolve ingredients, uniformly distribute powder particles and other additives, evaporate quickly, and not kill anyone or destroy the environment during these three stages. Fast drying is an important part of the solvent's job. The drying speed of a cast tape is the determining factor in casting speed. A slip that can dry in 10 minutes can be cast twice as fast as a slip that requires 20 minutes to dry, and it can thus produce twice the acreage of tape. A major reason some manufacturers hesitate to switch to aqueous-based tape casting is the reduced drying rate and the reduced production capacity that would be a result of the switch. The organic solvents mentioned, as well as others, evaporate much more rapidly than water. There are other organic solvents, such as 1,1,2 methyl pyrrolidinone, that are considered much less hazardous to human health and to the environment but do not volatilize rapidly enough to be considered useful in the manufacture of tape.

With drying rate so closely tied to production rate, it may seem best to use only the most volatile of the solvents and to evaporate them as quickly as possible. This choice, however, has more drawbacks than benefits in most cases. Drying speed must be viewed with the big picture in mind. There are often serious technical repercussions associated with any extreme measures. For example, consider a single-solvent system using methyl ethyl ketone (MEK). While the MEK dries quickly, allowing rapid production, it runs a high risk of drying prematurely and causing a "skin" to appear in the slip reservoir. This translates into streaks or furrows along the length of the cast. The rapid drying may also cause the slip to form lumps along the downstream edge of the doctor blade, causing streaks and occasionally letting a lump or "dumpling" disengage from the blade into the tape itself. These are problems that can be addressed mechanically, and they will be discussed later. There are other possible results of drying too quickly that cannot be mechanically avoided.

The "skin" formation on a drying cast is, for the most part, unavoidable, but it should be delayed as long as possible. A pure MEK slip may skin too early, causing a state very similar to "case hardening" in spray drying, where the dry (or drier) outer layer inhibits the drying of the interior. This is very pronounced when casting tapes on the order of 0.010" to 0.080" and thicker. The fast-evaporating solvent in these cases actually slows down the drying of the body as a whole. Adding some 95% ethanol slows down the evaporation process, delaying the formation of the solvent-depleted region and thus allowing the overall tape drying to take place more quickly. Another option chosen in some of the published literature has been to insert a small quantity of a less volatile solvent for the sole purpose of retarding skin formation. While referred to historically as a homogenizer, the additive (cyclohexanone) would be much more appropriately called a skin retarder, although some have reported that it aids dispersion.[23,28] This drying phenomenon will be discussed in more depth in the section on drying. Suffice it to say at this point that there are good reasons to use highly volatile solvents, and good reasons to slow down the drying of highly volatile solvents. A very efficient solvent system will be a balance between the extremes.

Some solvent systems that have proven effective over the years have been:

methyl ethyl ketone (MEK)[11]
MEK/95% ethanol[24,26,27,28,29,30]
MEK/anhydrous ethanol[25]
xylenes/95% ethanol
xylenes/anhydrous ethanol[18]
MEK/toluene[29,31]
MEK/acetone[32]
toluene[33]
1,1,1 trichloroethane (TCE)
TCE/anhydrous ethanol
TCE/95% ethanol[27,34]
TCE/MEK/ethanol
TCE/acetone
toluene/95% ethanol
MEK/95% ethanol/toluene
MEK/methanol/butanol[35]
MIBK/methanol

toluene/ethanol/cyclohexanone
MEK/95% ethanol/cyclohexanone
MEK/anhydrous ethanol/cyclohexanone
MEK/anhydrous ethanol/xylenes/cyclohexanone
butanol/isopropanol/xylenes/nitropropane[21]
many, many more. . .

As you can see, each of these solvents brings some advantages to the table and, since most of them are fully miscible if not soluble in each other, there are few interactions to deter the blending of beneficial properties. When the final choices of solvents are made, the usual criteria used for choosing which ones stay and which ones go are materials cost and time-to-market. Anhydrous ethanol is not a common choice for use with nonhydrophobic powders due to its high cost as compared to reagent grade 95% ethanol, and the benefits of a water-free slip do not often justify the higher cost. Ethanol and isopropanol are possibly the most common solvents chosen for a solvent-based cast. These solvents have a very good balance between drying rate, cost, safety and environmental considerations, miscibility in other solvents, and dissolving power for additives.

Azeotropes, mentioned earlier, are an interesting subject for debate. An azeotrope, or azeotropic mixture, is a blend of liquids that essentially acts as one. The azeotrope combines the dissolving capabilities of all of the solvents yet evaporates as a single liquid, retaining stoichiometry (or relative concentration) during the drying process. Some mixtures of solvents, such as ethanol/MEK, are azeotropic over a range of concentrations (> 46 wt% ethanol) whereas other mixtures, such as ethanol/toluene, are azeotropic at essentially a single point (68 wt% ethanol).[36] Some experienced practitioners in the field swear by azeotropic mixtures and would never consider using a nonazeotropic mixture. Our experience, however, has led to the belief that whether a solvent mixture is azeotropic or not makes little, if any, difference in the final tape. The relative concentrations of the solvents has a large effect on slip performance, but there is no drastic change in properties or characteristics when the solvent combination becomes azeotropic. As an example, xylenes and 95% ethanol is listed as a good solvent mixture and has no azeotrope. An interesting side note on this subject is that for some solvent blends, the solvent system becomes azeotropic part way through the drying cycle. (For example, an MEK/ethanol mixture at a 60/40 ratio is non-azeotropic. During the drying process, the more volatile MEK will evaporate more quickly than the ethanol, causing the component ratio to shift into a higher ethanol concentration. When the concentration of ethanol reaches 46 wt% or better as compared to the MEK, the solvent mixture is in the azeotropic range and then evaporates as a single liquid.)

Another crucial factor in choosing a solvent is the extent of regulation surrounding its use. Over the years, close scrutiny has fallen onto solvents such as xylenes, MEK, toluene, benzene, and 1,1,1 trichloroethane. Some of these solvents, as well as many others, have been banned for industrial use in certain states or countries, whereas others have been regulated stringently enough to make their use cost-prohibitive. Carcinogenicity is a major point of debate for these solvents, along with other long-term effects on the human body. As promised, this book will not delve too deeply into who's right or who's wrong, but the extent of debate and the changing regulations imposed upon the use and disposal of these solvents and their vapors can have a significant effect on the cost of using them. As stated above, it is strongly recommended (if not required by law) that every person exposed to industrial chemicals be familiar with proper handling techniques, storage requirements, and safety protocol surrounding the use of these chemicals.

While the term *solvent* is usually used in this field to describe organic solvents, it should be kept in mind that water still holds the title of "The Universal Solvent." Water is a nonorganic solvent that can be used by itself or in conjunction with organic solvents to produce viable tape casting slips. Since the main advantages of water are in the areas of cost and safety, there are not many published works combining organic solvents and water. The addition of an organic solvent defeats the purpose of using water at all.

Purely aqueous systems, however, have been looked at quite extensively for a number of years. Aqueous slip processing has a great deal of history behind it in fields such as drain casting, pressure casting, extrusion, and spray drying. To date, the main drawback in the use of an aqueous solvent system has been either its inability to dissolve the "right" binders or our lack of knowledge as to what the "right" binders are. The other demands placed on the solvent (cost, drying rate, environment and safety, chemical solubility) do not seem to be as big a problem. Binders of the type necessary for good green tape characteristics, however, have not been found in the "water soluble" category without some drastic side effects. The binder problems faced when working with water as the sole solvent will be covered in the aqueous processing section of this book, as will the ways around these problems.

The last requirement of the tape casting solvent is one that doesn't often need to be taken into account and wasn't mentioned previously. The solvent must not adversely affect the powder. Most of the organic solvents, and water for that matter, are fairly nonreactive with the vast majority of powders to be cast. Exceptions do exist, however. AlN substrates, for example, cannot be tape cast without accounting for the $AlN–H_2O$ interaction. Other water-sensitive and/or water-soluble powders have been cast that required totally anhydrous solvent systems such as salts, iron (Fe) powder, and powders with special organic coatings. These considerations, when applicable, simply limit the field of choices for solvents in the slip. MEK, xylenes, and anhydrous ethanol are common choices for totally anhydrous processing. Keep in mind, though, that an anhydrous solvent may not stay fully anhydrous once the storage container is opened.

Homogenizers

A homogenizer is an agent that works to make a system uniform throughout. The idea of homogenizers in tape casting has been around for a while. The term *homogenizer* as it appears in the literature typically refers to the organic solvent cyclohexanone. The term *homogenizer* was associated with this solvent due to legal and political issues at the time of its naming.[37] A more appropriate title would be "skin retarder." In small quantities, the cyclohexanone helps to avoid skin formation during drying.[23] Being typically the less volatile solvent in the system, it works to keep the top surface, or drying surface, of the tape liquid so that solvent in the body of the tape can more readily diffuse to the top face. Perhaps, in a roundabout way, it works to homogenize the concentration of solvent throughout the tape by retarding the formation of a solvent-depleted zone (skin) on the drying face. Since just about all of the organic solvents listed previously, and others, are fully miscible or soluble in each other, the need for a homogenizing additive rarely, if ever, exists. This does not mean that cyclohexanone is useless, just that its purpose doesn't fit its title.

Cyclohexanone is not always used in small quantities for a specific purpose. One published formulation uses cyclohexanone as the main solvent along with smaller additions of toluene and ethanol.[38] The retarding of skin formation is an important aspect to be considered in building a tape casting slip, but it can usually, if not always, be accomplished by correctly assembling and balancing the other slip components.

SURFACTANTS: What the heck are they?

Rich: A surfactant is usually a chemical to aid in wetting or modifying the particle surface.

Eric: Surfactant, Surface Active Agent. It *should* mean anything that gets on the particle surface and causes something to happen that normally wouldn't.

Rich: Why do you have to make a big deal out of everything? Just use the accepted meaning of the word.

Eric: I suppose I'm just contrary by nature. If a surfactant aids the wetting of a particle surface, which results in a new layer between primary particles, which in turn holds the particles apart and in suspension, then the chemical (which is an active agent on the particle surface) is a dispersant and maybe even a deflocculant. The word *surfactant* should describe dispersants, deflocculants, wetting agents, flattening agents (sometimes), flocculants, and many others.

Rich: Let's just avoid the use of *surfactant* and describe the actual function of the chemical in the slip system. Is that good enough for you?

Eric: That works for me—semantic propriety leads to clear communication. Let's look into these "surfactants" by treating each category as different additives.

The compromise is made. Let's explore the topic.

2.3 SURFACTANTS

Surfactants have been covered in just about every book on ceramic processing ever written. This is not surprising, since the role they play in ceramic processing is enormously important. They are the key to dispersion, wetting, high density, porosity, deflocculation, slip stability, green strength (in some processes), and just about any other measurable property in a formed ceramic shape. For those who already know what a surfactant is, please be sure to read this section very carefully, since we may not use the word *surfactant* the same way you do.

SURFACTANT is a contraction for 'SURFace ACTive AgeNT'. It is an additive that actively modifies (or coats) the particle surface to impart a desired characteristic, such as lower surface charge, higher surface charge, high/low surface energy, or specific surface chemistry. We use the word *surfactant* to describe any additive that causes the particles to do something that they normally wouldn't do by means of modifying (or coating) either a part of or all of the exposed surface of the particle. In this definition, just about everything we put into the tape casting slip works as a surfactant to some extent, with the possible exception of the plasticizers. This being the case, we will try not to use the word *surfactant* at all, but describe each additive according to its intended purpose.

2.3.1 Deflocculants/Dispersants

A deflocculant is an additive that works in the system to keep particles apart. The role of a deflocculant in a tape casting slip is fivefold: (1) to separate or hold separate the primary particles so the binder can coat them individually,[39] (2) to increase solids loading in the powder suspension in order to maintain moderate viscosities after binder addition, (3) to decrease the amount of solvent in the powder suspension in order to save money on solvents, (4) to decrease the amount of solvent in the powder suspension in order to dry the slip faster and with less shrinkage, (5) to burn out cleanly prior to sintering in order not to contaminate the final fired part.

When individual powder particles (primary particles) are in proximity, they often have a tendency to form loosely bound groups called *flocs*. The attractive force and binding force can be contributed electrostatically, by van der Waals forces, or by a number of other interparticulate forces. In a fluid suspension of particles, particularly when the powder has little affinity for wetting, flocs can form to lower the free energy of the suspension by reducing the solid–liquid interface area. We won't go very deeply into particle–particle attractive forces and DLVO theory in this book—it has been covered very well in many ceramics processing books by other authors[40,41] and doesn't need to be repeated again here. Regardless of where the particle groups come from or why they exist, they are detrimental to the properties of the cast tape and must be broken up.

A large family of additives has been found over the years that helps to get and/or keep particles away from each other. These additives are called deflocculants. Some confusion crops up very easily at this point, since many people, and many texts, use the terms *deflocculant* and *dispersant* interchangeably.[42] It is easy to see why the words are mixed together so frequently when you consider the result in the slip body. When a slip is deflocculated, the fluid (solvent, vehicle) is a continuous phase that fully surrounds each primary particle. A fully deflocculated slip typically displays a lower viscosity due to the particle mobility offered by the fluid interparticulate layer. It can be said about such slips that the particles are well dispersed.

The purpose of the dispersant is to disperse primary particles and to hold them in a homogeneous suspension. The dispersant additive can accomplish this action by two methods, steric hindrance and ionic repulsion. The dispersant separates primary particles and holds them in suspension, allowing the vehicle to form a separating layer between them. A well-dispersed suspension typically displays a lower viscosity due to the particle mobility offered by the fluid interparticulate layer. These slips are usually well deflocculated. Any surface chemist or scientist well versed in slip rheology will tell you that this definition is an exceedingly simplistic one. We agree. We are only going to look at the purposes and results of dispersants and deflocculants because, in the processing of tape casting slips, the fully deflocculated and well-dispersed slip is only the first step. This is only one of the additives in the system.

Deflocculation and dispersion, while two distinctly different needs, are usually accomplished with a single additive. As you might have seen, the same mechanisms that force the particles apart (or prohibit their coming together) also hold them in suspension. *Ionic repulsion* is the term used to describe the charging of particle surfaces so that they magnetically repel one another. In water systems, this can often be accomplished simply by controlling the pH. This surface charging can also be accomplished by introducing soluble polyelectrolytes, which coat the particle surfaces and provide (or negate) the surface charge. *Steric hindrance* is the term used to describe the separation of particles by putting a coating on one particle that will physically prohibit another particle from coming into contact with it. Organic chain molecules are often used in both aqueous and nonaqueous systems to accomplish this particle–particle separation. A list of common dispersants/deflocculants can be found in Table 2.1. It is worthy of note that most of the dispersants for use in organic solvents contain blends of various fatty acids and esters.

There are a number of reasons why a tape casting slip must be well deflocculated. First, the powder being added to the slip is usually in the form of soft agglomerates, particularly with high-surface-area powders. The groups of particles either in agglomerates or flocs tend to trap air in the interstitial space between the primary particles. If the particle group is not broken up, the trapped air will cause trouble in deairing later in the process, bubbles in the green tape, or unwanted porosity in the fired product. Second, when the polymeric binder is introduced, it will envelop the particle group instead of the individual particles. This action, which we call zipper bag theory, makes the particle group a permanent group throughout the rest of the process. zipper bag theory will be covered in the processing section of this book. Third, a well deflocculated slip will settle to a much higher packed-bed density than a partially flocculated (or coagulated) slip.[43] If there is any doubt about this, take 25 sheets of copy paper and shred them into 1/4" × 1/4" flat pieces, putting them into a container. Take another 25 sheets and crumble them into loose balls, putting them into an identical container. It should be easily seen that the small individual particles of paper form a much denser stack than the loosely formed "flocs" of paper. This higher packed-bed density leads to higher green bulk density in the tape, higher GOOD density, and higher fired density of the end product.[42] Fourthly, one of the determinants of a reliable and repeatable final product is the homogeneity of the slip and the resulting homogeneity of the green tape. Flocs or agglomerates in the slip cause inhomogeneity in the green tape by introducing local high-porosity regions within the interstitial void space of the particle group. The deflocculant acts against particle grouping, thus increasing homogeneity in the slip, tape, and final part.[44]

Table 2.1

Some Reported Dispersants/Deflocculants used for Tape Casting

Polyisobutylene[48]	pH adjustments
Linoleic acid[48]	Sodium silicate
Oleic acid	Dibutyl amine
Citric acid	Substituted imidazolines[47]
Stearic acid	Sulfanates
Lanolin fatty acids[47]	Aliphatic hydrocarbons[47]
Salts of polyacrylic acids	2-amino-2-methyl-1-propanol[47]
Salts of methacrylic acids	Polyethylene glycol[47]
Blown menhaden fish oil	Polyvinyl butyral[47]
Corn oil	Sodium sulfosuccinates[47]
Safflower oil	Ethoxylate[47]
Linseed oil	Phosphate ester[47]
Glycerol trioleate	Glycerol tristearate
Synthetic waxy esters	Many proprietary chemicals

In comparison, slip casting or drain casting is usually performed using a slightly flocculated slip. The slight flocculation of the slip allows some disruption of the particle packing against the walls of the gypsum mold which allows water, the usual solvent for slip casting, to pass through the packed bed. Since the particle packing of the slip-cast part is not as dense as it could be, the part can form at a faster rate. Slip casting relies upon flocs to create a lower-density packing structure that allows water to pass through it at a moderate rate. In our lab, we slip-cast a number of parts using a slightly flocculated slip and found that a 1/8" wall was formed in approximately 10 minutes. Casting the same part using a fully deflocculated slip required well over 45 minutes to build up the 1/8" wall thickness. The denser packed bed formed by the fully deflocculated slip did not allow water to pass through as rapidly. As a follow-up to that experiment, the fully deflocculated slip displayed less firing shrinkage due to the higher packed-bed density. In tape casting, however, we do not require that the packed bed be permeable by the solvent. The goal for packed-bed density is that it be as dense as possible. Of course, there are always exceptions if you want porosity in your fired part. We will talk about that kind of specialty application in Chapter 6 on applications.

One of the measurable effects of using a deflocculant is a higher packed-bed density. This phenomenon has been used to determine the proper dispersant level (remember that the deflocculant and dispersant are usually the same additive). The level or extent of deflocculation can be measured by a number of techniques, ranging from microelectrophoresis[41] to optical microscopy to laser scattering. Targeted variables of these tests include zeta potential, isoelectric point (IEP), and equivalent spherical diameter. If the time is available, however, we suggest doing a settling experiment. Since one of the main reasons for deflocculating the slip is to form a densely packed bed of particles, it logically follows that a screening experiment should be performed with the densest packed bed as the goal.

Tormey[45] described an experiment in which different concentrations of dispersant were used to disperse otherwise identical suspensions of powder and solvent. Based on that screening procedure, an experimental technique was developed to observe the extent of deflocculation in powder suspensions.

The Dispersion Study

6 batches

1. Dissolve (0 g, 0.5 g, 1.0 g, 1.5 g, 2.0 g, 2.5 g) dispersant in 200 g of the desired solvent (or solvent mixture).
2. Add the solution to a small mill jar approximately one-third filled with media.
3. Add 100 g of the desired powder.
4. Roll for four hours.
5. Add 10 ml of the resulting slurry to a 10 ml graduated cylinder and seal.
6. Allow to sit undisturbed for several days.
7. When all settling has been completed, note the packed-bed height as well as the clarity of the supernatant liquid.

The packed-bed height will decrease sharply with small additions of deflocculant but will reach a plateau beyond which no further decrease is seen (see Figure 2.3). (Note: For some deflocculants the plateau is not seen until higher concentrations are used.)

The density of the packed bed displays the extent of deflocculation. The dual action of dispersion and deflocculation can be observed as the higher additive concentrations take longer to finish settling. The plateau is reached when the deflocculant concentration is adequate to fully deflocculate the powder. Additional information can be gathered from this test by noting the clarity of the supernatant liquid. At higher dispersant concentrations, the supernatant liquid may be cloudy, since some of the particles stay in suspension. The use of a

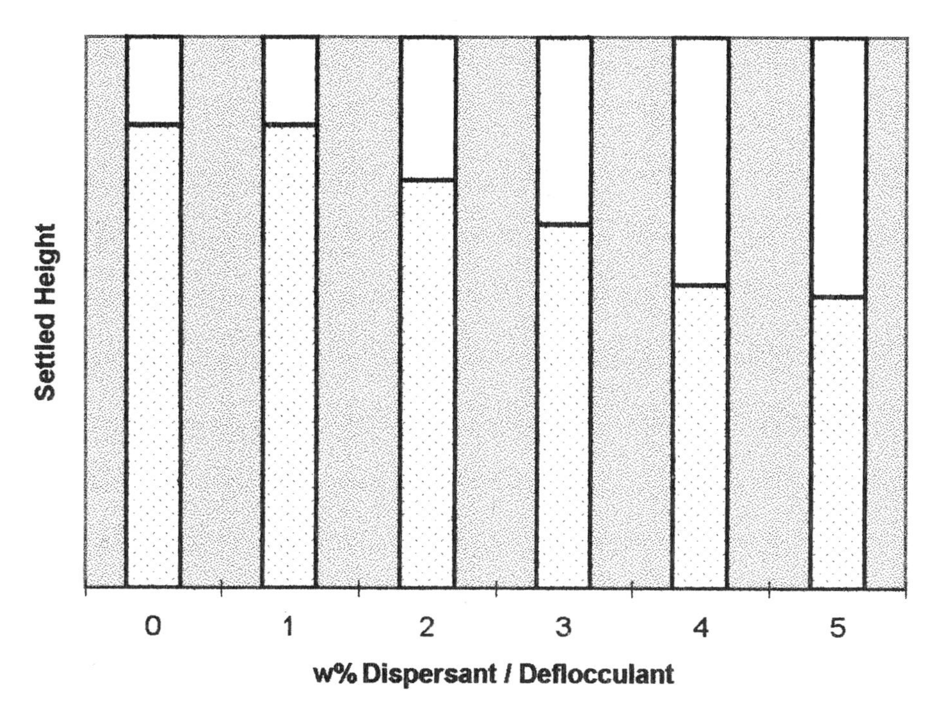

Fig. 2.3 The packed-bed height will decrease sharply with small additions of deflocculant but will reach a plateau beyond which no further decrease is seen.

colored deflocculant such as menhaden fish oil will also display the point at which excess deflocculant stays in solution rather than attaching to the particle surface. Low fish oil concentrations display clear supernatant liquids, which increase in amber color as the fish oil concentration increases.

The dispersion study described introduces another step in the deflocculation process, the ball mill. The milling action of the ball mill (or jar mill) mechanically separates primary particles. This will be covered in more detail in Chapter 3 on materials processing, but it requires some attention at this point.

Dispersion/deflocculation of a powder in a fluid vehicle takes place in three primary stages.[46] When the powder is first put in proximity to the fluid, the particle surfaces (or agglomerates) get coated with the fluid. Stage I of suspension is the wetting of the particle surface. After the particle surface is wetted, soft agglomerates held together by electrostatic forces and flocs held together by van der Waals forces need to be separated. Stage II of suspension is the mechanical separation of primary particles.[47] After the completion of Stage II, the particles are evenly distributed throughout the fluid. Stage III of suspension consists of protecting and stabilizing what was accomplished. Stage III is described as prohibiting flocs from reforming in the slurry. While the surfactant (as a wetting agent) may aid Stage I, the primary benefit of the surfactant (as a deflocculant) is seen in Stage III. From a big-picture viewpoint, most of the deflocculants used in tape casting should be called antiflocculants, since the separation of flocs into primary particles is really done by mechanical means, while the resistance to reflocculation is provided by the surfactant.

Most of the dispersants/deflocculants used in solvent-based (organic solvent–based) tape casting are of the steric hindrance type. Also called steric stabilization, this is a mechanism in which one part of a chain molecule attaches to the particle surface while the rest remains dissolved in the fluid vehicle. The "tail" that is in solution acts as a boundary layer, physically prohibiting particle–particle contact. The lower dielectric constant typically associated with organic solvents makes steric stabilization the logical choice for a dispersion mechanism. Water, which is discussed in its own chapter, has a higher dielectric constant and offers both steric and ionic dispersion options.

The ball mill, jar mill, or mill jar is a sealable ceramic container that is charged with pieces of ceramic, the balls. Some processes may benefit from using metal balls and/or metal containers. Cost considerations may even lead to the use of plastic containers. The purpose of the ball mill in tape casting is to thoroughly mix all of the additives into a homogenous suspension. During slip processing, the actual breaking up of agglomerates and flocs is accomplished by the turbulent mixing of the slip inside the mill. Particle groups are also broken up by physically "mashing" them between the falling ball and the wall of the mill jar. Once the particles are separated, the dispersant/deflocculant is able to coat the entire particle surface and maintain the mechanically induced separation. Thinking about the three stages of deflocculation, the wetting stage is not complete until all of the primary particles are forced apart during Stage II. The charging and function of the dispersion mill will be covered again, in much more detail, Chapter 3.

Dispersion mechanisms have been fairly well researched in nonaqueous solvent systems, primarily with oxides. While, again, the main particle separation mechanism has been found to be steric hindrance, some have concluded that both steric and ionic mechanisms are active in the organic solvent–based suspension.[48] The ionic repulsion dispersion mechanism may come into play in organic solvent systems with certain additives but, since tape casting slips generally contain a high solids loading and the particles are in close proximity to each other, steric hindrance is still an important stabilization method.[47] The extent of surface charge interaction would, of course, be dictated by the solvents chosen, specifically their dielectric constant and degree of polarity. The balance of steric and ionic stabilization forces, while important and worthy of scrutiny, gets overshadowed quickly when the binder(s) and plasticizer(s) are added to the system.

Remember that the main purpose of the deflocculant is to hold the particles apart with a continuous fluid layer between them. Many get caught up in analyzing the dispersion/suspension and lose sight of the big picture. Take care to keep in mind that the deflocculant is only the first additive in the system and that the ball milling continues to provide mechanical separation after the binder is introduced to the slip. The polymeric binder, in some cases, overlaps the dispersant/deflocculant function[49] and can act as a more powerful dispersant/deflocculant or, more directly, antiflocculant than the deflocculant itself. Mikeska and Cannon even list poly(vinyl butyral) among dispersants.[47] Does this make the binder a surfactant? MacKinnon and Blum[50] mention a decrease in viscosity seen with the addition of a binder into a fine particle suspension and a reverse effect in a suspension of coarser, broad PSD particles. We have seen viscosity increases and decreases with fine particles, coarse particles, wide and narrow distributions seeming to follow the dispersant/binder combination. While there is really no way to specifically confirm or deny this statement, we have both seen adequate evidence to feel that, out of the organic additives remaining in the dry tape, the binder has the largest effect on slurry viscosity and particle separation by far. This dual function of the binder as a deflocculant/dispersant is covered in a bit more depth in the organic interactions section of this chapter.

2.3.1.1 Fish Oil

Over the years, menhaden fish oil (MFO) has been one of the most common dispersants used in tape casting. Initially sought out as a "leveling agent" by Harold Stetson's group at AT&T, it was seen to work well as a deflocculant with just about any powder it was put in with, acting as a fully steric hindrance deflocculant[51] in one alumina suspension analyzed. The high regard for its deflocculating ability is due to the large variety of fatty acid esters it contains. Fish oil is a natural additive. It contains, among other things, 20 or more different types of fatty acids, ranging from linoleic to palmitic to myristic to stearic (see Table 2.2). Fish oil is primarily, if not entirely, a "steric hindrance" deflocculant, though the dielectric nature of the

Table 2.2

Typical Fatty Acid Composition of Menhaden Fish Oil

(Determined by gas-liquid chromatography—Karlshamns)

Fatty Acid	Carbon Atoms	C=C Double Bonds	Percent
Myristic	14	0	9.3
Myristoleic	14	1	0.3
Palmitic	16	0	19.3
Palmitoleic	16	1	11.9
Hexadecadienoic	16	2	1.7
Hexadecatrienoic	16	3	1.7
Hexadecatetranoic	16	4	1.6
Margaric	17	0	1.2
Stearic	18	0	3.4
Oleic	18	1	11.4
Linoleic	18	2	1.3
Linolenic	18	3	1.3
Octadecatetraenoic	18	4	3.3
Arachidic	20	0	0.2
Gadoleic	20	1	1.9
Eicosatetraenoic	20	4	2.3
Eicosapentaenoic	20	5	14.7
Erucic	22	1	0.5

Note: The remaining percentage of fatty acids exists in very small weight fractions too numerous to list.

solvent can significantly affect this. It is a mixture of long-chain fatty acids that attach to the particle surface and trail out into the fluid vehicle. There are other dispersants—some natural, some synthetic, some polymeric—that give lower viscosities in certain systems, but none (so far) have been as effective across the board as menhaden fish oil. Both authors are supporters of the use of MFO, primarily due to its ability to ignore minor fluctuations in powder surface chemistry and bulk powder characteristics. Shanefield mentions increased lot-to-lot uniformity as compared to more refined additives, owing to the large variety of oxidized esters available in the MFO.[52] Other authors have noted difficulty in dispersing certain powders in specific solvents such as ethanol, xylenes, or toluene.[42] MFO has been seen repeatably to be an excellent dispersant for alumina in toluene/ethanol or xylenes/ethanol systems.

There are many varieties of MFO to choose from. The standard practice is to use an oxidized (blown) version. It was found that, unaltered, MFO is a rather poor deflocculant for oxide systems, while the oxidized blown menhaden fish oil gives the needed deflocculation. It has been communicated by one manufacturer[53] that blowing the oil develops an active polar ester group. Oxidation raises the free fatty acid content, which aids the complete wetting of small particles. The polar ester group(s) of the fatty acid can attach strongly to surface (-OH) groups[54] on oxide powders in single or multiple locations. The extent of oxidation for MFO is measured by viscosity and is reported in terms such as "Blown Z-3" for medium viscosity and "Blown Z-8," which is more oxidized and much higher viscosity. It has been noticed as well that the undesirable fishy odor decreases substantially with the extent of oxidation. One potential drawback to the use of MFO is the fact that, over time, it will oxidize by itself. This is not generally considered a major concern, however, since the process takes a long time and does not significantly affect its ability to deflocculate a powder. The authors have compared MFO lots differing in production times by as much as 10 years and achieved identical results. We have seen, however, with one fine-grained alumina, a significant change in dispersing effect between new and 15-year-old MFO. Since most manufacturers do not store a 10-year supply of chemicals, self-oxidation in storage is generally not a problem. Another potential concern centers on slight compositional variations across years and/or seasons.[55] Since MFO is a natural additive, gained from the menhaden fish, the fatty acid concentrations shift slightly with different seasons and changing environmental conditions. This also has not been seen to affect its role in the tape casting systems observed. The wide range of functionality found with MFO is most likely due to the broad spectrum of fatty acid types it contains.

Along with being a forgiving and functional deflocculant, MFO has other benefits to the tape casting process. These other benefits, which will be discussed in the "Organic Interactions" section of this chapter, include: carrier release, Type II plasticizer (lubricant), and lamination aid.

2.3.1.2 Phosphate Ester (a.k.a. Organic Phosphate Ester Acid)

Much has been reported in recent years about phosphate ester for use as a deflocculant/dispersant for tape casting slips. Phosphate ester is a very powerful dispersant for many oxide powders, including $BaTiO_3$, Al_2O_3, and TiO_2. Phosphate ester is soluble in either water or a number of organic solvents and is described as an anionic surfactant in polar liquids.[56] Overall conclusions about this additive are that it functions both as an ionic repulsion and steric hindrance deflocculant.

Experimentation by the authors has shown that, in MEK/ethanol–alumina systems, the phosphate ester works extremely well and allows relatively high solids loading. Overall batch loading showed that, as a dispersant/deflocculant, the phosphate ester could be considered a direct substitute for MFO and an improvement over some polymeric deflocculants. One potential drawback associated with the phosphate ester is the contamination of the ceramic body by residual phosphorus.[55] For electronic applications, phosphorus is a contributor to dielectric conductivity and is thus undesirable. Residual phosphorus in the form of P_2O_5 is also a "sponge" for MgO at elevated temperatures and can lead to unwanted liquid phase sintering in high-alumina bodies. Since MgO is used as a grain growth inhibitor for pure alumina by ensuring solid-state sintering, the P_2O_5–MgO glass phase leads to undue grain growth and a resulting higher grain size and corresponding lower density.[57] For doped aluminas, other oxides and nonelectronic applications, the residual phosphorus may not be as great a concern.

With the big picture in mind once again, the great benefits of lower viscosity and/or higher solids loadings achievable with phosphate ester tend to be lost with the addition of the tape casting binder. In one comparative test, while the dispersion mill showed a much lower viscosity at higher solids loading as compared to MFO, additional solvent was required with the binder addition due to the thickening action of the binder. The two slips, varying only in dispersant type (phosphate ester vs. menhaden fish oil blown Z-3), had similar viscosities at identical solids loadings after all ingredients were added. This is certainly not to say that one additive is more or less useful than another, but quite the contrary: they essentially do the same thing in the system as a whole as it pertains to preparing the particles to meet the binder.

That, after all, is the bottom line as it pertains to the deflocculant/dispersant. The purpose of dispersion and deflocculation—breaking apart the particles, wetting the particles, holding a fluid layer between primary particles, and whatever else goes on in the dispersion mill—is to prepare the particles to meet the binder and behave in a beneficial manner while the binder does its work. The separation of particles is to allow the binder to attach to each particle separately instead of surrounding groups of particles. Many, many additives exist that can fulfill this need for separation. Literally hundreds of chemicals have been looked at for this purpose over the years. While we have only looked in depth at two of these additives, some of the referenced works list many others to try.[41,47,52,58,59]

BINDERS: They hold the system together, yet hold the system apart.

Every ceramic formation process uses binders to hold the system together, to hold the ceramic particles in relative position to each other. Clays use the organic material inherent in the natural clay product. Some casting formulations take advantage of electrostatic forces, van der Waals attractive forces, etc. Some use very stiff binders to retain shape, like polyacrylic dispersant/binders in slip casting. Some use binders that are mushy when wet but harden during drying.

Tape casting is the only process that requires the binder, and the body as a whole, to remain very flexible even after drying. To some extent, spray drying needs to generate a powder that has a "soft" binder to facilitate pressing (isostatic, dry, ram or extrusion), but the final shape needs to be fairly stiff to withstand handling or machining.

This section talks about binding agents for particles; organics to hold individual particles together. Just about all of the binders used have the ability to be plasticized or are fairly soft to begin with. The dry tape needs to be flexible, formable, punchable, etc.

Enough of the hype. Let's explore the topic.

2.4 BINDERS

The binder or binders used in the production of green ceramic tape (or tape of any material) are probably the most important processing additives of the system. The binder supplies the network that holds the entire chemical system together for further processing. Essentially, green ceramic tape is a polymer matrix impregnated with a large amount of ceramic (or other) material. In actuality, the powder is encased in the polymeric resin during the slip preparation process, but the resulting green product of the casting process is a continuous-phase resin that surrounds entrapped particles. It would be accurate to call the end product "loaded polymer tape."

Being the only continuous phase in the green tape, the binder has the greatest effect on such green tape properties as strength, flexibility, plasticity, laminatability, durability, toughness, printability, and smoothness. From the standpoint of further green processing, the tape can be looked upon as pure binder, since every exposed surface of the tape product is binder. Two metaphors (similes actually) can be used to talk about tape in the green state.

(1) Green tape is like a sponge in which the sponge material is a polymeric resin, and instead of holes, the normal void space is filled with ceramic particles (or other particles). This metaphor falls short in that the ceramic is not a continuous phase like the void space in a sponge.

(2) Green tape is a matrix of ceramic or metallic particles joined to each other, point to point, with polymer fastening ties.

These models are expanded in the following section on plasticizers. This section should be considered inseparable from Section 2.5 due to the extensive interaction between binders and plasticizers in the system. It should be kept in mind that regardless of the model used, porosity is a third major phase in the green tape body. This green tape porosity gives rise to the concept of theoretical tape density.

During the drying process, the fluid vehicle (solvent) evaporates from the system, leaving void space behind. Some of the void space disappears as the body contracts (drying shrinkage), while some of the volume previously occupied by the solvent remains as porosity. Residual porosity in the tape can be quantified by comparing the measured green bulk density to the calculated theoretical green bulk density. Residual porosity contributes to further processing variables, such as compressibility, strength, printability (ink adhesion), and shrinkage. Porosity in the tape will be covered in more detail in later chapters.

As mentioned in the previous section on solvents, the binder is typically the second ingredient chosen when building a tape casting slip, preceded only by the powder to be cast. The priority of the binder over the other ingredients stems from its power to determine the green tape character. Factors taken into account when choosing a tape casting binder include solubility, viscosity, cost, strength, T_g or ability to modify T_g, firing atmosphere of the powder, ash residue, burnout temperature, and by-products.

Considering the final part to be manufactured (the whole reason why the tape is cast), consideration must be given to the firing atmosphere. It would not serve to use an excellent air burnoff binder for casting silicon carbide if the SiC will be fired in a reducing atmosphere. This is an especially important consideration when casting metal powders or other nonoxide powders.

Many different binders have been used through the years for making adequate and even excellent green ceramic tapes. The majority of these binders fall into two main families: polyvinyls (vinyl) and polyacrylates (acrylic). Each of these families has strengths and weaknesses. The main difference between them is the burnoff/removal characteristics in different atmospheres. Across the board, the one thing that all tape casting binders have in common is their "film forming" capability. It is mandatory that a tape casting binder be a film former almost by definition, since the green ceramic tape is merely a highly loaded film. Typically these film formers are long-chain polymers or precursors (monomers or emulsion particles) that become long-chain polymers during drying.

The polymer chain length is not necessarily fixed for each polymer. The desired polymer may be available in a number of variations defined by polymer chain length, categorized by molecular weight. Lower molecular weight polymers are shorter in length (fewer *-mer* units per chain), resulting in a lower viscosity when dissolved. The resulting slip will tend to have a lower viscosity, thereby allowing a higher solids loading. This is often a trade-off, however, since the lower molecular weight polymer will tend to yield a weaker tape requiring a larger amount of the low molecular weight binder. Some hold to the practice of lowering the binder molecular weight with decreasing particle size and increasing chain length with larger particles.

2.4.1 Vinyl

Vinyl binders include many different groups and are well known in other industries, including the consumer markets. Most people are familiar with at least one vinyl polymer in everyday life—(poly)vinyl chloride, or PVC. PVC is commonly used for water and sewer pipes as well as some lawn furniture. Many other processing fields, including the food and textile industries, use (poly)vinyl alcohol (PVA) on a regular basis. Both of these binders have also been used to make ceramic tape. The most commonly reported polyvinyl resin in the tape casting field, however, is (poly)vinyl butyral, which is also referred to as PVB or by the common tradename Butvar®. PVB is often used in the textile industry as a protective coating on fabric, and it is an ingredient in the production of safety glass. Butvar® is a registered trademark of Solutia, Inc., a large supplier of the chemical. Polyvinyl butyral has become a very common tape casting binder due to a number of beneficial characteristics.

PVA is generally considered a "water-based" binder. Since some PVAs are water soluble, the strong desire to steer clear of flammable organic solvents usually steers development into an aqueous solvent system. PVC has also been tried as a tape casting binder but is not a commonly used ingredient. PVB (polyvinyl butyral) has been extensively reported as a tape casting binder and has been studied at great length with respect to its role in the tape casting system.

The vinyl family is typically used for powder systems to be fired in an oxidizing atmosphere, either air or oxygen. Wet hydrogen, wet nitrogen, and cracked ammonia atmospheres can also provide good removal of the binder. The atmospheric limitations of the binder are due mainly to the method of decomposition of the polymer. The vinyl polymer burns at an elevated temperature and thus typically requires oxygen in the atmosphere for full burnoff. The typical remnant of the vinyl polymer after firing is carbon. This is most often removed by the presence of oxygen to form CO or CO_2, but it can also be removed, as mentioned, by the presence of water vapor or hydrogen and to a lesser degree by the presence of nitrogen. Residual carbon (ash) is typical of vinyl decomposition in nonoxidizing atmospheres or atmospheres that do not supply reaction components for the carbon. When choosing a vinyl binder for a tape casting process, pay attention to the volatile components typical to burnoff in the desired firing atmosphere. Specifically, these burnoff components may include: carbon monoxide, butyraldehyde, water, and small amounts of various acid compounds. Shih et. al. have studied the pyrolysis of Butvar® in detail and have published their results for your reference.[60,61]

Plasticizers are quite varied for the vinyl family, depending on the type of vinyl polymer. Many plasticizers are known for PVB through decades of development trials. Plasticizers for PVA tend to be more limited in number. PVC similarly has few reported plasticizing agents.[62] The function and types of plasticizers are addressed in detail in Section 2.5. It should be mentioned at this point, however, that atmospheric humidity can act as a plasticizer for PVA, which can have very major repercussions during post-cast processing.

2.4.2 Acrylics

Acrylic binders or "acrylates" are also widely used, long-chain-film-forming binders for tape casting. Some versions of acrylic polymers used for tape casting binders are listed in Table 2.3. Some of the benefits of acrylic polymers include: cost, cleaner removal in neutral or reducing atmospheres, strength, solubility, and ability to modify T_g. Some of these benefits, such as solubility and T_g modification, are shared with members of the vinyl family. Acrylic binders are quite different from the vinyl family, however, in decomposition mechanism. While the vinyl binder burns (oxidizes), acrylic polymers disassemble and evaporate. Many practitioners in the field refer to acrylic decomposition as the "unzipping" of the acrylic chain. This decomposition mechanism facilitates the removal of the acrylic binder in reducing or neutral (inert) atmospheres with little "ash" or carbon residue. For powders that require reducing or inert atmospheres, such as high-purity SiC, metal powders, or AlN, acrylic binders are often chosen.

Acrylic binders are also popular in tapes that are sintered in vacuum for their burnout, or "decomposition" qualities.

Many acrylic polymers, specifically (poly)methyl methacrylate and (poly)ethyl methacrylate, show great strength at low concentrations in the green tape. This benefit also contributes to binder removal later in the process, since lower binder content leads to lower binder residue. Acrylic binders are also plasticized or made flexible by a wide variety of additives, which gives you the freedom to choose a plasticizer that burns out cleanly. The acrylic binders, when used as dissolved polymers, can be made flexible by both Type I and Type II plasticizers. These plasticizers are addressed in detail in Section 2.5. We have also seen the methacrylates exhibit behavior quite the reverse of "common sense." It has been seen in more than one solvent system, with different powder chemistries, that more binder can lead to denser tape. This will be discussed in Section 4.3.

2.4.3 Cellulose

Many cellulose polymers have been explored through the years as tape casting binders. As with PVA, much of the tape casting development with cellulose polymers has been with aqueous solvents, since many of the cellulose polymers are water soluble. The prospect of an aqueous tape casting slurry is so attractive for safety and environmental reasons, that little work has been done with cellulose in organic solvents for tape casting. Aqueous tape casting using cellulosic binders has been reported many times.[63,64,65] Various types of cellulose binders reported in tape casting use are listed in Table 2.3.

The drawbacks of the cellulose family of binders are large enough that very few actually use a cellulose binder for green tape production. One of the biggest drawbacks of cellulose-type binders is their thickening properties. Methyl cellulose and ethyl cellulose (with various side groups) form very viscous aqueous solutions that are quite detrimental to tape-cast processing. Pumping of the slip, filtering, and flow under the doctor blade all require a moderately fluid slurry

Table 2.3

Reported Binders Used for Tape Casting

VINYL	**CELLULOSE**
Polyvinyl alcohol	Cellulose acetate—Butyrate
Polyvinyl butyral	Nitrocellulose
Polyvinyl chloride	Methyl cellulose
Vinyl chloride—Acetate	Ethyl cellulose
	Hydroxyethyl cellulose
ACRYLIC	Hydroxypropyl methyl cellulose
Polyacrylate esters	
Polymethyl methacrylate	
Polyethyl methacrylate	
OTHER	**AQUEOUS BINDERS**
Petroleum resins	Polyvinyl alcohol
Polyethylene	Celluloses:
Ethylene oxide polymer	Ethyl, methyl, hydroxyethyl
Polypropylene carbonate	hydroxypropyl methyl, above
Polytetrafluoroetylene (PTFE)	Emulsions of:
Poly-alpha-methyl styrene	Acrylics, latex
Poly isobutylene	Polypropylene carbonate
Atactic poly(propylene)/Poly(butene)	PVB, waxes
Polyurethane	

which, with cellulose binders, typically requires excess solvent (water), which lowers the slurry solids loading. Low solids loading increases drying shrinkage, cracking tendency, and tape porosity, and it slows the drying process. One system was reported to have an 8:1 wet-to-dry thickness ratio.[65]

Another drawback often seen when using cellulose binders is difficulty in deairing. Cellulose-bound slips tend to exhibit evidence of large bubble populations in green tapes. Many methods have been tried to limit air entrapment in water-based cellulose slips, ranging from the simple to the extreme. These will be covered in Chapter 7. The last major drawback to cellulose binders is shared with PVA. The cellulose family has a limited number of potential plasticizers, and those available tend to be Type II plasticizers. Extensive use of Type II plasticizers has drawbacks that are discussed in Section 2.5. The most readily available Type I plasticizer desired for the water-soluble cellulose binder is water in the form of atmospheric humidity. The water-soluble cellulose binder, like the water-soluble PVA binder, will change in binding properties with changes in atmospheric humidity. All of these factors tend to make other binders, even nonaqueous binders, more attractive for use in tape casting. Binders are available for use in aqueous systems that are not water soluble but instead are in emulsion form. These water-based emulsion binders will be addressed in Chapter 7.

2.4.4 Other Binders

Many other binders have been used for tape casting, as can be seen in Table 2.3. Still others have been used and not published. The choice of binder for a tape casting system, as mentioned earlier in this section, is perhaps the most important choice to make when designing a tape casting slip. Solubility is an issue when solvent choices are limited by powder reactivity, health, safety, or environmental boundaries. Clean burnout is an issue when bisque and sintering atmosphere choices are limited by powder or other component needs. Strength and toughness characteristics are an issue when downstream handling, shaping, or forming is strenuous.

One specialty binder is worthy of note due to its claimed burnout characteristics. Polypropylene carbonate is a film-forming polymer that decomposes fully in any atmosphere, including vacuum, leaving no binder residue in the bisque or sintered part. Potential drawbacks to this polymer include a lower strength per weight as compared to other polymers in green tapes and water as a combustion product. Water-sensitive powders may be formed into green tape without detriment to the powder chemistry only to be hydrated or chemically damaged during firing by the water that is formed by polymer combustion. One of the great benefits of this polymer is its ability to be plasticized by its own monomer, propylene carbonate, which also evaporates cleanly with no ash residue. Proper firing through the binder removal stage can also avoid combustion of the polymer, allowing evaporative, rather than combustive, removal of the polymer and thus yielding an ash-free and water-free system through casting and firing.[66]

PLASTICIZERS: They make things plastic... or do they?

What do the following things have to do with tape casting?

Bathroom tiles	Garbage bags
Rubber bands	Olive oil
Tree branches	Pottery clay
Extension cords	Toothpaste
Notebook paper	Chewing gum

To find out the connection between tape casting and bathroom tiles, read...

2.5 PLASTICIZERS

This section of the book will help to explain the function of the binder in a green ceramic tape and help you to visualize ways of enhancing binder performance. Most tape casting formulations include at least one additive referred to as a "plasticizer." The word *plasticizer* is used very loosely in the tape casting field; it can refer to just about anything that makes the tape more bendable. This section will try to explain the different types of binder modifications or tape modifications that are available, along with the mechanism used to modify the green tape.

The purpose of the plasticizer in a green tape is to enable the tape to be bent, or to "give," without cracking. Most of the polymeric binders used for forming tapes will form a relatively strong, stiff, and brittle sheet if no plasticizer is used. Later forming and assembly needs require the ability to punch, cut, roll, or laminate the dry tape. The role of the plasticizer is to allow a hole to be punched in the tape without shattering or cracking the tape, a 500-foot roll of tape to be stored in inventory for later use, a three-by-four inch substrate to be blanked out of a sheet of tape without cracking the surrounding tape, and so on. Plasticizers work either on or around the binder polymer chains to allow motion inside the tape matrix without breaking the matrix itself.

Two words need to be defined before continuing this discussion. Most engineers and/or ceramists use the words *plasticity* (being plastic) and *flexibility* (being flexible) to mean very different things. Flexible simply means "able to be bent without breaking." A sheet of paper has flexibility. Plasticity is used to describe the ability to permanently deform. Pottery clay has plasticity. Bending a sheet of paper does not change the structure of the paper. Laying the paper on a flat surface will make the paper flat again in its original form. Rolling a sheet of clay into a tube will permanently deform the clay to some extent. Laying the tube on a flat surface may result in a half tube, but the clay will not easily revert to a flat sheet again. As another example, extension cords and tree branches in the wind show flexibility, while toothpaste and chewing gum display plasticity. While these are obviously different characteristics of a material, it is possible for one item to have both properties. Plastic bags, for example, show great flexibility and can return to their original shape. If enough tensile force is placed on the bag, however, the bag will stretch and permanently deform. This permanent deformation is referred to as *plastic deformation.*

The word *plasticizer* in tape casting is loosely used to describe any additive that imparts either flexibility or plasticity to the green tape. The plasticizer has been defined by the International Union of Pure and Applied Chemistry in this manner: "A plasticizer or softener is a substance or material used to increase its (the binder's) flexibility, workability, or distensibility."[67] There are advantages and disadvantages of both of these characteristics in green tape. The next section introduces a visualization tool or model that will be most helpful in understanding the interactions between binder polymer chains and different types of plasticizers. This tool will be referred to as the "mosaic tile model."

The green tape can be viewed, as mentioned in the previous section, as "a matrix of ceramic particles joined to each other, point to point, with polymer fastening ties." For a down-to-earth visual model, picture the 1" × 1" mosaic tile pattern on your bathroom wall border, your bathroom floor, or perhaps a shower enclosure. For ease of installation, the tile manufacturer provides these small tiles evenly spaced and attached on a metal wire mesh backing in various sheet sizes (12" × 12", 12" × 24", etc.). If these sheets are stacked on top of each other to form a 24-inch-high stack, with vertical ties being added to attach the separate sheets together, we have a rough idea of what the green tape looks like. This "mosaic tile model" will be used to help explain the role plasticizers play in the tape.

NOTE: In the mosaic tile model, please note that the particle-particle ties provided by the binder not only hold the particles together but also hold them apart. This particle separation function of the binder is a "steric hindrance" or "steric stabilization" mechanism that aids and sometimes overshadows the dispersing property of the dispersant. It was mentioned in the Surfactants section that the binder is sometimes a more powerful dispersant than the dispersant itself. The mosaic tile model provides a good picture of the dispersing power of some binders.

Two distinctly different mechanisms can be used to plasticize a green tape. Confusion arises quickly, however, since the end result of these different mechanisms looks very similar in most tests. The mosaic tile model will be useful in displaying the two different mechanisms. As we left the model, we have a 12" wide, 24" long, and 24" high block of 1" × 1" ceramic tiles bound together with metal wires. Bending this stack around a pipe would require not only a fair amount of force, but it would also require the outer layer of wires to stretch, since the ceramic tiles need to move farther apart. The first type of plasticizer is a chemical that softens the polymer chains between particles, allowing them to stretch more easily. This "softening" plasticizer is referred to in this book as a Type I plasticizer, a T_g modifier, an internal plasticizer, or a binder solvent. In the mosaic tile model, it would be equivalent to replacing all of the metal wires with rubber bands. With these new rubber band connections, the 1" × 1" tiles can more easily move apart to bend around the pipe, yet straightening the block allows the tiles to be pulled back to their original position. This Type I plasticizer is described in depth later in this section.

With the mosaic tile model now bound with rubber bands, another issue limiting the ability to bend is the friction inside the block. Tiles scrape on tiles, rubber bands scrape on rubber bands, rubber bands scrape on tiles. Enough force on the block will make the block bend, and it will bend to a certain extent without cracking, but the stiffness of the block may be too high. A second type of plasticizer is available that imparts plasticity (ability to permanently deform) to the green tape matrix. The mosaic tile model will bend with less effort if olive oil is poured over the whole system to reduce the friction. The oil will act as a lubricant between the tiles inside the matrix, allowing easier deformation of the matrix as a whole. Consider also that, since the lubricant is added to the tape casting slip before all of the binder ties are formed, prior to cross-linking, the physical presence of the lubricant prevents some of the ties from forming.[67,68] This preventative action allows some particles to be moved inside the body of the tape without a great deal of force compelling them to return to the original position, thus decreasing rigidity. The mosaic tile model placed on top of the pipe will now bend under its own weight around the pipe. This additive that lubricates the internal matrix of the green tape is referred to in this book as Type II plasticizer, body plasticizer, external plasticizer, or lubricant. This Type II plasticizer is also discussed in more detail later in this section. Note that some authors refer to both Type I and Type II plasticizers as external plasticizers.[68] Due to this semantic confusion, we will avoid using the terms internal plasticizer and external plasticizer.

As mentioned earlier, many of the tests that can be used to analyze green tape will not reveal the difference between the effects of the different plasticizers. For example, bending the tape around a known radius may show the extent of elongation before cracking (see section 5.1.4), but the tape may avoid cracking either by stretching of the "rubber bands" or by permanent motion of the matrix. Further analysis would be needed to determine the reason for not cracking.

It can be seen using the mosaic tile model that relative changes in Type I and Type II plasticizer additions will cause logical and predictable changes in tape flexibility and plasticity that can be charted by or even read from a standard stress-strain diagram.

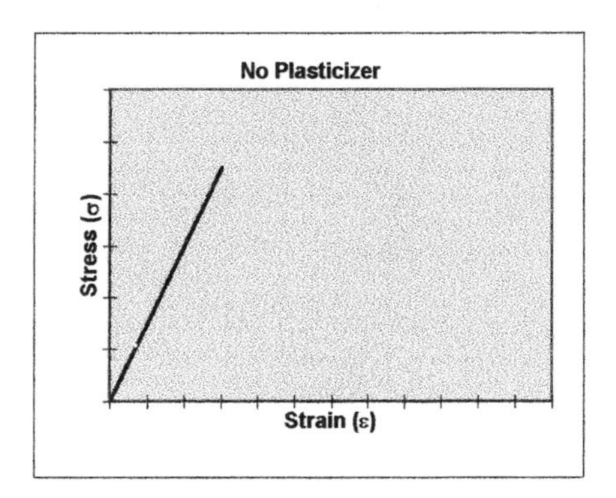

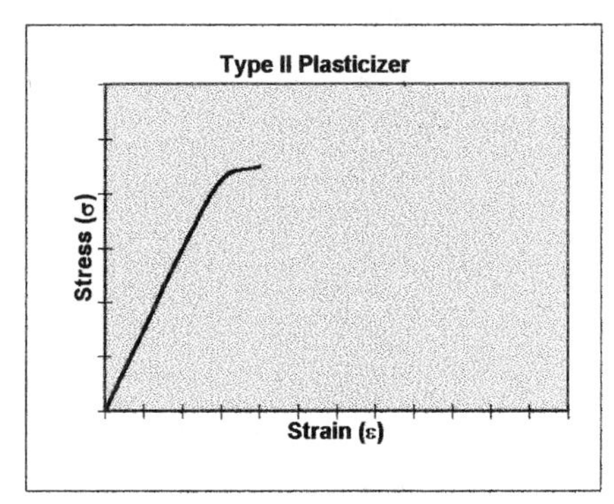

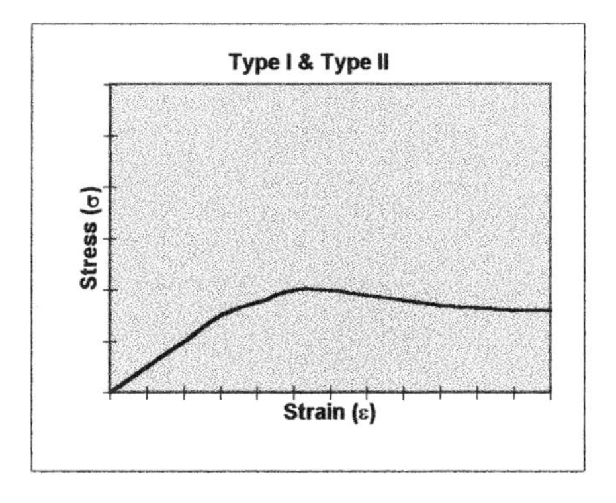

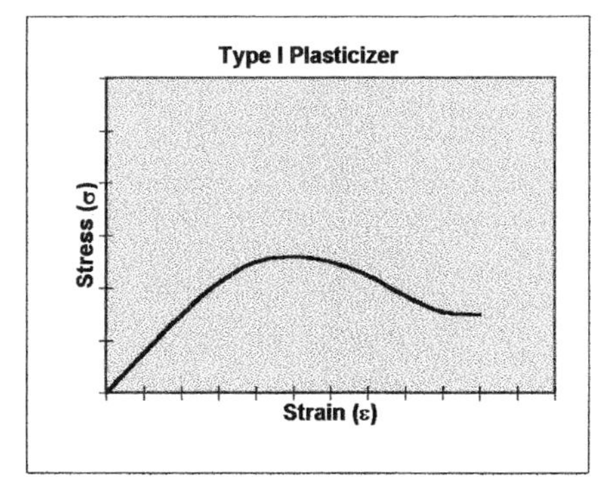

Figure 2.4 Effect of plasticizer type on stress-strain properties.

Because the addition of a Type I plasticizer will soften the polymer, lowering the T_g of the binder, the slope of the stress/strain curve would be expected to be less than that of the unmodified binder. Adding a Type II plasticizer to the system allows motion of the matrix and increased yield before failure. The pure Type II plasticizer, by comparison, will display a lower strain to failure,[69] because the binder remains stiff without a lubricant to soften the polymer. Note in addition to these behaviors that modification of the polymer T_g lowers the yield stress of the polymer in all cases. It should also be noted that a behavior similar to dislocation pinning occurs during tape elongation, which can be seen in the reversing curvature of the stress/strain diagram at higher strains. This pinning increases the slope of the diagram and is seen just prior to failure in most tapes. The use of both Type I and Type II plasticizers not only allows polymer yield but also facilitates permanent motion within the matrix and bypasses some of the pre-fracture stiffening of the polymer. Tapes made with high concentrations of both Type I and Type II plasticizers are very claylike in consistency and are easily deformed and molded.

A listing of commonly used plasticizers is included in Table 2.4. We have arranged these plasticizers in the following categories: phthalates, glycols, and other. The sections that follow will describe these plasticizers in more detail.

2.5.1 Type I Plasticizers

Type I plasticizers are used to soften the binder polymer chains, allowing them to stretch or deflect under an applied force. These additives can be accurately described as T_g modifiers or binder solvents.[70] T_g is a symbol that stands for glass transition temperature. While crystalline solids have a distinct change from solid to liquid, the binder polymer in the dry tape is a noncrystalline solid, having a gentle transformation from solid to liquid. While crystalline solids freeze and melt, the polymer matrix softens more and more until it is considered a liquid. In order to describe the softening characteristics of a polymer (or glass), a temperature is calculated

Table 2.4

Reported Plasticizers Used for Tape Casting

Phthalates	**Glycols**
n-Butyl (dibutyl)	(poly)Ethylene
Dioctyl	Polyalkylene
Butyl benzyl	(poly)Propylene
Mixed esters	Triethylene
Dimethyl	Dipropylglycol dibenzoate
Other	
Ethyltoluene sulfonamides	
Glycerine (Glycerol)	
Tri-n-butyl phosphate	
Butyl stearate	
Methyl abietate	
Tricresyl phosphate	
Propylene carbonate	
Water*	

* Humidity acts as a plasticizer for water soluble polymers.

to estimate an appropriate "melting point." This calculated temperature is called the liquid-solid transformation point or glass transition temperature (T_g). Nothing spectacular happens at this temperature, but it serves as a descriptive reference for the noncrystalline material.

Two ways in which the Type I plasticizer can modify the T_g of a polymer chain are by shortening the polymer chain length and by partially dissolving the polymer chain. Both of these mechanisms accomplish the same result, making the tape more flexible at a given temperature. As the T_g becomes lower, often well below room temperature, the polymer chain is better able to stretch or reorient itself without fracturing. An example of polymer chain shortening is seen in the case of polypropylene carbonate modified by propylene carbonate. The propylene carbonate shortens the number of units in the long-chain polymer, thus weakening the polymer and allowing it to yield. In comparison, many members of the phthalate family are good solvents for the binder[72] and dissolve the binder, allowing it to yield—thus the label *binder solvent*. Essentially, the only difference between a Type I plasticizer and the solvent vehicle is the volatility or rate of evaporation.[67,71] Adding a Type I plasticizer, provided that it is compatible with the binder polymer, should increase flexibility and extensibility and decrease the strength of the green tape.[73] Essentially, the use of a Type I plasticizer creates an extended leather-hard state by incorporating a very slow-drying solvent into the slip.

Large amounts of a Type I plasticizer will make the tape very elastic and can make the surface quite tacky. Excessive amounts of a Type I plasticizer will cause "blocking," the immediate adhesion of one layer to another, due to the liquid-like character imparted to the tape. Excessive use of a Type I plasticizer has major drawbacks in that the adhesion of the polymer to the carrier surface increases while the yield stress of the polymer decreases. A tape that has been over-plasticized by a Type I plasticizer will stretch rather than release from the carrier surface.[69]

The Type I plasticizer is a "binder solvent" similar to the solvent vehicle, as mentioned previously. The most notable difference between the "solvent" and the Type I plasticizer is the rate of evaporation (volatility). Among the Type I plasticizers there are some fairly volatile chemicals, along with some low-volatility chemicals. Propylene carbonate, for example, is volatile enough to be listed as a solvent in some trade literature. Dioctyl phthalate is also a fairly volatile ingredient, as is n-butyl phthalate to a lesser degree. These ingredients are not nearly as volatile as the solvent vehicle, but they evaporate quickly enough to warrant some protective measures. Tapes using these ingredients should be stored in a sealed container to avoid brittleness in the aged tape. Other Type I plasticizers, such as butyl benzyl phthalate or mixed alkyl phthalate, are very stable (less volatile) in the dry tape and do not require protective measures. Most solvent-based tape casting includes a Type I plasticizer.

2.5.2 Type II Plasticizers

The Type II plasticizer works as a lubricant in the tape matrix.[69] Some, but far from all, solvent-based tapes include a Type II plasticizer. The Type II plasticizer works between the polymer chains, not only allowing them better mobility within the dry tape, but also preventing some of the "cross-linking" between chains.[68] Referring back to the model, a Type II plasticizer prevents some of the rubber bands from forming, weakening the matrix and allowing better mobility within the matrix. This increased mobility results in larger strain to failure but also decreases the matrix's yield stress. On the production floor, this results in less cracking of the tape during handling, but it creates a higher probability of plastic (irreversible) deformation, which will show up in the fired part.

The ability to plastically deform is present in tapes even without Type II plasticizers. Under heat, pressure, or tensile stress, any plastic will yield. The Type II plasticizer lowers the yield stress and increases strain to failure. The lower yield stress is beneficial for nonplanar applications of tape, such as wrappings or coatings for brazing applications. The low yield stress also works in concert with the plasticity in planar applications such as multilayered electronics. The plastic deformation of a green sheet is necessary when lamination to a textured surface is desired. One sheet of tape is often required to adhere fully to the tape sheet below it around printed metallization lines or pads. The texture imparted to the base layer of tape by printed metallization requires the second layer of tape to conform to a given shape. The ability to deform, or flow, into surface topography can increase bond strength between layers, avoid void space around metallization lines or pads, and decrease delamination populations in further processing. (See Figure 2.5.) Due to this beneficial effect, Type II plasticizers are sometimes referred to as "lamination aids."

The increased yield-to-fracture and low yield point enhanced by an added lubricant (Type II plasticizer) can help to prevent cracking in thicker green tapes during drying. As will be discussed in more detail in Section 4.3, stresses develop during the drying process due to the one-side-drying configuration of tape casting. With all of the solvent evaporating from the top of the tape, and the bottom of the tape restrained by the carrier surface, internal stresses develop that can lead to curling, cracking, or premature release from the carrier due to lateral stress. The ability to yield under the stresses of drying provided by the Type II plasticizer can help to prevent cracking in tapes up to 0.150 in. thick by allowing the matrix to "give" under the drying stress and yield a flat, crack-free tape. Higher percentage elongation of the top surface is also necessary to roll thicker tapes for storage. The addition of a Type II plasticizer will increase the tendency of the tape to roll up onto the take-up reel rather than to release from the carrier. Take care to notice, however, that during rolling only the top surface stretches. This may induce warpage or unidirectional camber in a fired part, stemming from the permanent (plastic) deformation of the matrix during rolling.

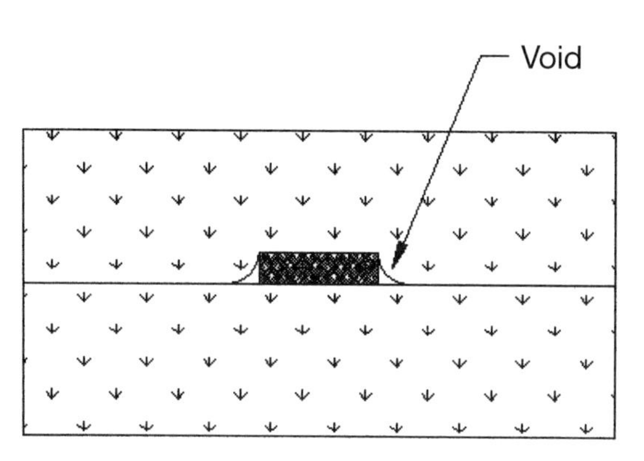

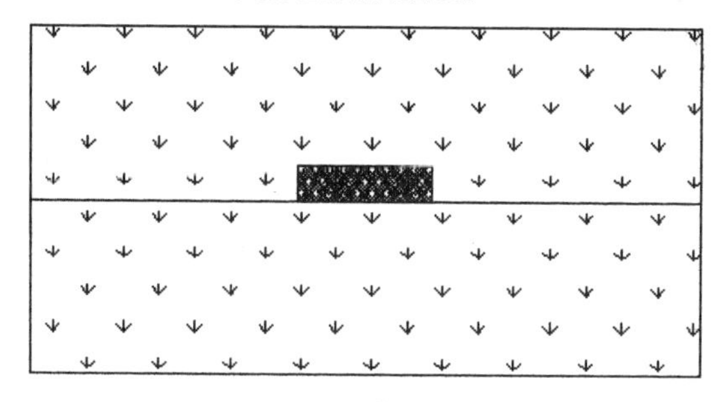

Fig. 2.5 The ability to deform, or flow, into surface topography can increase bond strength between layers, avoid void space around metallization lines or pads, and decrease delamination populations in further processing.

One benefit of the Type II plasticizer that was not found in the published literature is its lubricating benefit in the tape casting slip before drying. Tape casting slips in general are very highly loaded suspensions and tend to be pseudoplastic (shear thinning) in nature. When this pseudoplasticity is extreme, generally in the case of very fine particles and/or high–surface area powders, the addition of a lubricant aids mobility inside the fluid prior to drying, lowering the extent of shear thinning behavior by decreasing the low-shear viscosity. Hayashi does mention lubrication of the particle surface to avoid dilatant (shear thickening) behavior in the slip.[68] This lubricating effect of the Type II plasticizer can aid flow under the doctor blade and avoid streaks or grooves in the green tape, since pseudoplasticity, thixotropy, and even dilatancy are caused by binding forces in the fluid. The lubricant lives up to its title in such cases and lubricates the binding interfaces in the slip, allowing lower yield stress in the fluid.

The last benefit of a Type II plasticizer also stems from the lubricating function. The fact that this type of additive interferes with the binder's binding qualities allows the lubricant to act as a release agent. Incorporating a Type II plasticizer into the tape formulation is seen in some cases to enhance the release characteristics from the casting carrier. The Type II plasticizer acts as a release layer on the bottom side of the tape, essentially lubricating the interface between the tape and the polymer, steel, glass, or granite casting surface. This action can be beneficial for single-layered products but can inhibit the lamination process in multilayered products by continuing to provide a release layer between green sheets.

The use of a Type II plasticizer should lower the yield stress of a tape-cast sheet, increasing its ability to deform under its own weight (that is, it is floppier). Extensive use of a Type II plasticizer will result in a moldable ceramic sheet that acts much like a clay body in its ability to stretch and flow. Excessive use of a Type II plasticizer will lower the yield stress and lower the strength of the tape to an unusable level by preventing too many of the polymer ties from forming. The tape becomes unusable when the yield strength becomes less than the tensile stress imparted by either the force of gravity or the stresses imparted during tape handling.

Other than lowering the yield stress of the tape and facilitating plastic deformation, which can be beneficial or detrimental, another potential drawback to Type II plasticizers is phase separation or migration. As a lubricant, this additive has little or no chemical reaction with the other components in the system. The result of this inert behavior can have great benefits but can also generate nonuniformity. Some Type II plasticizer/binder combinations are not very compatible, where the lubricant has little affinity for the binder molecule and does not remain homogeneously distributed in the tape matrix. Some tapes will display "raindrop" patterns of Type II plasticizer on the top surface after the plasticizer is squeezed out of the matrix during drying shrinkage. Excessive use of a Type II plasticizer may display this phase separation even in compatible systems. The worst-case scenario of this phase separation results in a stiff tape with a very oily surface. Even when the additives are compatible, concentrations of Type II plasticizer can sometimes be worked out of the tape during repeated bending or through prolonged storage, forming a sticky or slippery surface.

2.6 ORGANIC INTERACTIONS

Entire books, as well as many articles, have been written about dispersion, including the effect of the dielectric constant of the fluid, particle surface charge, zeta-potential, steric forces, etc. The main thing that hinders such a book from describing all of the interactions in a tape casting slip is the vast number of potential interactions and the interdependency of one interaction on other interactions. At this stage of the technology, the interactions have been noted and empirically addressed but are not theoretically explained in their entirety. Even those interactions that have been examined and explained are only explained for very specific and controlled cases. Tape casting is a young technology that is mature enough to apply to new ideas and new applications but is not fully understood in all aspects. This section will explore the additive interactions seen in tape-cast processing. Solutions derived by experience and theories about the mechanisms at work are offered.

2.6.1 Binder / Dispersant Interactions

It was mentioned in the sections on surfactants and binders that the binder can act as a better dispersant than the "dispersant" does. The binder is intended to get between primary particles and to eventually (after drying) hold the individual particles together. The binder's action of getting between the particles also acts as a steric hindrance dispersion mechanism. In some cases, the dispersing action of the binder overshadows the dispersing action of the dispersant. It is characteristic of these cases that the viscosity of the slip will decrease after the binder is added. This can be deceiving, however, since it may take in excess of six milling hours for the viscosity to become noticeably lower. If the dispersing action of the binder is not as powerful as that of the dispersant, the slip viscosity is typically seen to increase due to the solution of the binder polymer. The interaction between the binder and dispersant depends upon a number of different variables, including type of binder, type of dispersant, powder chemistry, solvent choice, powder particle size distribution (PSD), powder surface area, type and amount of plasticizer, and possibly others.

One study of this interaction that we performed involved the use of polyvinyl butyral and two different dispersants, menhaden fish oil and phosphate ester. Batches were charged identically, with the exception of the dispersant. One batch was formulated with 4 wt% menhaden fish oil per powder weight, while the other was formulated with 2 wt% phosphate ester per powder weight. After dispersion milling, the MFO batch was very viscous, almost a paste (>20,000 cP), while the phosphate ester batch was extremely fluid (<100 cP). After identical plasticizer and binder additions, the slips were measured to have identical viscosities (approximately 2500 cP). In the MFO batch, the binder's dispersing action far overshadowed the fluidizing action of the MFO. Conversely, the phosphate ester was seen to be a much better dispersant/deflocculant than the MFO. The tapes cast from these slips were nominally identical, as were the fired densities and yields.

One factor that will be discussed in a more practical aspect in the next chapter is time. Many of the reactions and chemical interactions that occur in the preparation and processing of a tape casting slip require time. Some interactions progress to final state within a workable time frame, while others do not reach "end state" for an unreasonable (by manufacturing standards) amount of time. One of the binder/dispersant interactions that has been noticed progresses over a very long period of time.

Some binders have the habit of replacing some dispersants on the particle surface, displacing the dispersant back into the solvent vehicle. This has been most noticeable in the polyvinyl butyral-MFO combination. It is possible, however, that this interaction is most noticeable in the PVB-MFO combination because this has been one of the most widely used combinations for tape casting through the development of this technology. Over time, the PVB is driven to the same bonding sites as the MFO on the particle surface. This displacement puts a higher concentration of MFO back into the solvent vehicle and later into the interparticle spaces of the green tape or onto the surface of the dry tape. The interaction also decreases the concentration of interparticulate binder polymer in the system, essentially using up more binder on individual particle surfaces thus leaving less binder to attach one particle to another.

Referring back to the mosaic tile model in Section 2.5, over time some of the rubber bands from particle to particle wrap around the particles instead. This action displaces the menhaden fish oil which then acts like a Type II plasticizer to lubricate the system.[74] The end result of this interaction follows logically out of the mosaic tile model. Fewer rubber band ties between particles result in a weaker tape. More olive oil, or in this case fish oil, in the matrix results in greater deformation under its own weight and higher plastic deformation. As aging time increases with both PVB and MFO in proximity to the particle surface, the tape becomes weaker and floppier and can even discolor the drying surface of the tape.

In examining the extent of this displacement, one production level slip was aged in the ready-to-cast state for over five years, still in the ball mill. Normal production tapes were also stored in the dry, green tape form for this period. In the normal production process, the time between binder addition and casting ranged from 18 to 48 hours. It was often seen that the tapes cast after 18 hours were very slightly stiffer than those aged for a longer period. All of these tapes were strong, flexible, and defect free. After aging for five years, it was considered safe to assume that the dispersant displacement by the binder had progressed to its fullest extent. The mill was then rolled for a few hours to remix the slip. The tapes cast with the five-year-old slip were seen to exhibit point source and along-the-length cracking and were extremely weak, often breaking under their own weight. The five-year-old slip also yielded a tape that left behind a heavy residue on the carrier film. The displacement of dispersant by the binder left less binder to hold the system together, and that interparticulate binder which was left was very overlubricated.

It is essential at this stage to point out that this aging effect is a slip aging effect, not a tape aging effect. The tape stored for five years was similar to the as-cast condition. This detrimental interaction is seen when the slip is aged after the binder addition, but it is not seen in the green tape after drying. The practical solution for this potentially detrimental interaction is simply not to age the ready-to-cast slip. If the time between the binder addition and the tape casting is held reasonably constant, the extent of displacement will also be repeatable, as will the green tape properties.

Taking this line of reasoning one step further, there is an obvious reason to go through a dispersion milling step prior to the binder addition. This will be looked at again in a more practical manner in Chapter 3. The addition of the dispersant/deflocculant without the presence of the binder facilitates the coating of the primary particles with the chosen dispersant. The full coverage of the particle surface with the dispersant encourages the binder to stay where it belongs, between particles with each polymer chain attaching to multiple particles, not around an individual particle attaching in multiple locations on the same particle. When the dispersant and binder are added at the same time, there is a race for the particle surface that not only results in less binder between particles (it wraps around instead) but also results in more dispersant in the body of the tape (the binder used up the surface bonding sites). This is often a hidden occurrence due to another mechanism that we refer to as zipper bag theory. Mentioned briefly earlier, zipper bag theory addresses the tendency of the binder to coat (wrap around) a group of particles rather than each individual particle. The binder, added too soon, sees a multiparticle group as a single large particle and coats the group. This agglomerate has a lower surface area than the separate particles would, which results in more binder between particles.

The two actions, racing for the particle surfaces and coating particle groups, affect the interparticulate binder concentration or "free binder." *Free binder* is a term used to describe the quantity of binder not occupied by coating a particle surface. Of the two competing actions, zipper bag theory usually has a greater net effect, thereby increasing the free binder content of the tape even though a higher percentage of particle surface is coated with binder. This can hide the fact that the particle did not get full coverage with the dispersant. The detriments of this competition tend to show up in the dry tape in the form of lower GOOD and fired density,[75] tape and fired substrate surface roughness due to agglomerates, green and fired porosity, and irregular grain growth during sintering.

As can be seen, the binder interacts with the dispersant in some cases, competing for the particle surface. While only one firm example has been examined here, this is a normal occurrence for all binder-dispersant combinations, although the time required for this displacement may be so large as not to be noticed. As a rule of thumb, large polymers tend to replace smaller polymers on the particle surface.[76]

2.6.2 Binder / Plasticizer Interaction

The interaction of the binder additive and the plasticizer(s) is usually, if not always, very desirable. Much of the interaction between these two components has already been addressed in this chapter in either the binder section or the plasticizer section. For the most part, the only purpose for adding the plasticizer, especially the Type I plasticizer, is to interact with the binder. Referred to previously as a binder solvent, the Type I plasticizer's role in the system is to lower the T_g of the binder to increase the flexibility of the dry tape, extending the leather-hard stage to longer working times. This in teraction with the binder is the only reason that a Type I plasticizer is used.

The interaction between Type I plasticizer and binder, however, takes time. Adequate processing time must be allowed for the Type I plasticizer to modify the binder polymer and to increase the flexibility of the green tape. Insufficient reaction time will yield a stiff tape (high polymer T_g), which can lead to excessive curling or cracking in the tape during drying. It is generally recommended to allow the binder to be in the proximity of the Type I plasticizer for at least 12 hours before casting the slip. This interaction does not need to be in the presence of the powder to be cast and can be accomplished prior to addition to the mill. The practical reason for and application of this prereaction is described in Chapter 3.

The Type II plasticizer is also discussed in Section 2.5. The Type II plasticizer does not chemically react to any great extent with the binder or with the Type I plasticizer. The only real interaction between the Type II plasticizer (lubricant) and any other component(s) in the system is mechanical. As mentioned earlier, the lubricant physically inhibits some of the polymer ties from forming and facilitates permanent, plastic motion in both the fluid slip and the dried tape.

2.6.3 Dispersant/Plasticizer Interactions

The relationship between dispersants and plasticizers is minimal. Any interaction that may occur would be better termed multifunctionality rather than interaction. Some dispersants, typically fatty acid esters or other oils, have multiple functions within the green tape. As with most dispersants, only a percentage of the dispersant added is actually adsorbed on the particle surface. That portion which either is driven off the surface by the binder, or which never attached, can act as a lubricant in the system. The "excess" dispersant will act in exactly the same manner as would a Type II plasticizer, providing lubrication and plasticity and aiding in carrier release. This is unusual to see in dispersants that are solid in their pure form, but it is seen more regularly when using liquid dispersants such as glycerol trioleate, corn oil, or safflower oil. The Type II plasticizer gains its functionality by being a separate liquid phase in the dry tape. This plasticizing effect would logically not be seen in a tape in which the dispersant was not a liquid when dry.

2.6.4 Organic/Powder Interactions

In at least one instance, the OH content of the binder (in this case polyvinyl butyral) had a profound influence on the amount adsorbed onto aluminum oxide particles. [77] This in turn influenced the slip viscosity, which increased as the OH content increased. The increase in viscosity led to a decrease in the green bulk density, which ultimately led to an increase in firing shrinkage. Linear firing shrinkages ranging from 16 to 17.5% were reported for OH contents ranging from 16 to 23%. The moral of the story is: For applications that require shrinkage tolerances of 0.1%, the OH content of the PVB has to be tightly controlled.

CHAPTER 3

Materials Processing: Slip Preparation

MATERIAL PROCESSING: Let's make some slip!

Any process is limited by the materials you start with. Making a device from tape is limited by the quality of the tape. The quality of the tape is limited by the raw materials and how they are prepared into a slip.

The better your slip is, the better your tape *can* be, the easier it is to make high-quality, high-yield pieces. Making the best slip isn't the easiest thing in the world, so let's explore materials processing.

3.1 PRE-BATCHING POWDER PREPARATION

Often in ceramics processing, the incoming powders are not in the form needed for the final product, or they have undesirable properties for the chosen process. Incoming raw materials are often the main source of variation in a ceramic process. The tape casting process is no different from any other ceramic fabrication process in that regard. The removal of surface contamination, the control of surface water or degree of hydration, chemical surface treatments to promote coupling to the dispersant, surface chemical treatment to protect the surface from hydration, and the addition of specific dopants to promote sintering, all fall in the category of pre-batching powder preparation. An excellent example is the addition of magnesium oxide to aluminum oxide using a soluble salt such as magnesium nitrate dissolved in water and then drying and calcining to yield a uniform coating on the surface of the alumina particles. Several of these pre-batching procedures will be described in this section.

3.1.1 Powder Washing

Many ceramic powders processed by dry ball milling include an additive that prevents balling-up of the particles during the milling. These additives, generally called milling aids, are usually on the surface of the particles when the powder is delivered. A technique developed at the Western Electric Company[1] involves a washing process in which 95°C distilled water is poured through a bed of as-received alumina powder. The powder is then dried in an oven at 130°C for at least 18 hours. The powders are then added to the ball mill directly from the oven while still hot. It was found in that work that the washing procedure decreased the sodium concentration by a factor of 4 and improved the electrical resistivity of the resulting ceramic substrate by an order of magnitude. Similar types of washing procedures have been described to reduce the residual sodium and other alkali impurities from the surface of ferrite powders produced by chemical co-precipitation.[1] Large-scale washing and filtration of ceramic powders can be accomplished economically in continuous centrifugal filters or Funda-type[2] self-cleaning filters. The washing procedure only works when the milling aid is soluble. If an organic additive, which is insoluble in water, has been used, then a low-temperature calcination step may be needed. The temperature to use can be determined by thermogravimetric analysis (TGA).

3.1.2 Drying of Powders

Often the powders to be used in tape casting are very hygroscopic. They either pick up moisture readily from the atmosphere or were never properly dried and stored in the first place. Materials like this must be dried prior to batching in order to remove the physically adsorbed moisture from the particle surfaces and from between the particles. This water is sometimes referred to as free water, or water of hydration. Materials such as lithium aluminate are notorious for adsorbing moisture from the atmosphere. These materials are dried and stored in an oven at greater than 100°C, and they are added to the ball mill directly from the oven. In the development of the Western Electric process described in the previous section, it was thought that physically adsorbed water created a problem with viscosity variability and porosity in the dried cast tapes.[3] Therefore the powders were dried thoroughly and added to the mill directly from the oven. Since the time of that work, several papers have been published and much research has been done to determine the effect of water on organic solvent–based slurries.[4,5] Rives and Lee concluded that free water in an alumina slurry prepared with organic solvents such as toluene and tetrahydrofuran (THF) tended to have a large effect on the dispersion of the powder. In one case with the use of linolenic acid as a dispersant, the viscosity decreased with the addition of free water, and in another case with the use of polyisobutylene as a dispersant the viscosity increased with the addition of free water. In these cases, the free water in the solvent reacted with the alumina surface and formed a hydroxylated layer, which interacted differently with acidic or basic dispersants. Calcination of the alumina powders at temperatures above 900°C for long periods of time—four days in some cases—also had a pronounced effect on the viscosity, which increased because it lowered the degree of hydroxylation on the alumina powder surface. Hydroxyl groups on the alumina powder surface appear to be necessary to form a well-dispersed system for these dispersants. The fully dehydroxylated alumina powders could not be dispersed in either of the solvents. This leads us to the next section.

3.1.3 Surface Hydration

We believe that the use of surface hydration, that is, the heating of a powder in the presence of atmospheric humidity to form hydroxyl groups on the surface, is a critical procedural step in the processing of some oxide powders for use in tape casting. The use of temperatures greater than 100°C to "dry" powders actually does drive off some of the physically attached water, but it also drives the hydroxylation process to form a surface with a more uniform concentration of hydroxyl groups. It is very difficult to control the extent of surface hydration, so we try to push it as close to 100% as possible. The use of these hydroxylated powders directly from the oven yields more consistent results with respect to slip rheology in the tape casting slurry. This has been found and confirmed by other researchers for other ceramic processes.[6] The proposed mechanism for the effect caused by the hydroxyl groups on the surface of the powders is to promote a stronger chemical adsorption of the dispersant and therefore better rheological characteristics.[7] This has been observed with linolenic acid,[4] menhaden fish oil,[8,9] and phosphate ester.[10]

3.1.4 Chemical Surface Treatment

There are many instances where a beneficial effect is created by the use of a chemical surface treatment of the powders prior to batching. The use of silane derivatives has been reviewed in several articles.[11,12] This is a common treatment for the silicate fibers in epoxy-based printed wiring board technology. The silane organometallic compounds are commonly called "coupling agents," since they can build a bridge between the inorganic particle surfaces and the organic surfactants and binders in the tape casting slurry. The work described by Lindqvist, et al.[12] was applied to injection molding mixes, but we have seen the same effect of lower viscosity and higher solids loading in tape casting slurries of alumina.

Another use for silane treatment as well as for organics such as steric acid or octadecanoic acid has been to protect the surface of the particles from hydration so that they can be processed in a water-based system. Aluminum nitride, some of the ceramic superconductors, and materials like lithium aluminate fall into this category. The use of an organic "shield" to produce a monolayer of hydrophobic material on the surface has been reviewed in several articles.[13,14,15,16]

3.1.5 Pre-Batch Dopant Addition

There are cases where dopants are added to the major constituent prior to batching for tape casting. Some of these are in the form of soluble salts such as nitrates, which are premixed with the primary material and calcined to react uniformly over the entire surface area of the host powder. We prefer, when possible, to use a dopant that is soluble in one of the solvents used in the tape batch so that the uniform mixing is integral with the batching (or dispersion milling procedure). There are situations where this is not possible and where prereaction with the dopant is required. Adding very small quantities of dopants to a host material is an art in itself and is beyond the scope of this book. Usually the trick is to find a soluble salt that can then be dissolved in a liquid so that a uniform coating is formed on the surface of the major powder ingredient. One such trick we have used is the doping of alumina with a magnesium-based organic dispersant. This "kills two birds with one stone," because the magnesia promotes densification of the alumina without excessive grain growth while at the same time providing the necessary dispersion/deflocculation of the tape casting slurry.

3.2 DISPERSION MILLING

The first step in the preparation of a tape casting slip is dispersion milling in the presence of a dispersant/deflocculant in the solvent(s) of choice. The ball mill size is chosen according to the size of the batch to be processed. Ball mills range in processing volume from 0.3 liter to over 3000 liters. In the laboratory and in small-scale production, mills with volumes of 5 to 100 liters are common.

Other types of milling equipment have also been used for tape casting slurry preparation. For example, vibratory mills and shot-filled paint mills have been used. By far the most commonly used mills are standard rotary ball mills. The type of material used for both the mill and the grinding media is also an area of concern. When processing a high-purity powder that cannot stand to pick up mill impurities, one must work with a high-density, high-purity mill such as a high-alumina mill charged with high-alumina grinding balls or cylinders. Do not be fooled into thinking that "high-alumina" means 99% or better. In most cases a high-alumina mill means 85% alumina or better. The "high alumina" nomenclature was coined to distinguish the mill and media material from a porcelain material containing about 60 to 65% alumina. In some applications, the use of a high-density polyethylene jar is dictated to avoid inorganic contamination of the material being processed. These are available in many sizes ranging from one-half liter to large 10-liter carboys. Our experience has shown that the energy put into breaking down agglomerates is severely reduced by the use of these polyethylene jars. In some cases we actually had to risk impurity pickup by using an alumina mill in order to break down the agglomerates so that adequate green tape densities could be achieved and tape defects could be eliminated. The impact of ceramic media on a ceramic mill lining is lost when polyethylene mills are used.

Other mill choices include rubber linings or polyurethane linings. Both of these also reduce the energy put into the dispersion milling process. The choice of a synthetic liner would be dictated by the solvent(s) in the formulation and the degree of contamination that can be tolerated in the final formulation. Steel mills and grinding media are sometimes used for the milling of ferrite slips to control the impurity pickup and therefore the final chemistry. Media selection is not only predicated on the contamination problem but also on the specific gravity of the slip itself. We were very surprised in this regard when we used alumina media to mill a tungsten carbide slip. When the mill was opened, the media were actually floating! High-specific gravity slips must use high-density grinding media. The choice of grinding media shape is sometimes dictated by what is available from the equipment manufacturer. We have worked with spherical, cylindrical, and "natural" shaped flint stones, depending upon the powders to be processed. The usual grinding media mill charge used during dispersion milling ranges from one-third to one-half of the mill volume.

Once the mill size and material have been selected, the next step in the process is to load (or charge) the mill with the solvent(s), dispersant, wetting agent(s), and the powder(s) to be dispersed. The usual procedure would be to dissolve a predetermined amount of dispersant, based upon the settling tests described previously, in a weighed or volumetric amount of one of the solvents, either by stirring or by milling. The remaining weight or volume of solvents is then added to the ball mill. At this point in the process, the mill contains the grinding media, the solvent(s) and the dispersant. In some processes a wetting agent is also weighed and added to the mill. The preconditioned powder(s) are then, and only then, added to the mill. The mill is then closed with a sealing gasket to keep the liquid from leaking during the extended dispersion milling procedure. The choice of a sealing gasket can be simplified by using PVA gaskets for solvents such as MEK and acetone and Viton™ for alcohol-based solvents. Standard rubber gaskets should only be used for aqueous-based slips.

We define the solids loading in the dispersion mill as follows:

% Solids = Powder Weight / (Powder Weight + Solvent Weight)

Percent solids loading during dispersion milling can range from less than 20% to more than 90%, depending on the powder density, particle size, and dispersant effectiveness. We usually try to optimize the solids loading during this step, since any excess solvent will have to be removed during the drying process at a point downstream. A typical aluminum oxide dispersion mill with excellent dispersion would fall in the range of 70 to 80% solids loading. In many cases, the solids loading will have to be adjusted by trial-and-error techniques, since the solvent volume also affects the final viscosity after the binder and plasticizer(s) are added to the mix. The viscosity of the dispersion mill is not a primary concern, as long as deagglomeration occurs. Some dispersions will seem extremely fluid at a 75 wt% solids loading but will increase in viscosity when the binder is added. The highest possible solids loading in the dispersion mill is not important, since the binder will be added later.

As mentioned previously, the purpose of the dispersion milling procedure is to accomplish three things: (1) to break apart any agglomerated particles by the pounding and grinding action of the media, (2) to coat the broken-apart particles with a dispersant, which (3) keeps the particles apart by steric effects, electrostatic effects, or a combination of both. The primary objective in most, if not all, tape casting dispersion mills is *not* particle size reduction. There is bound to be a small amount of size reduction in all milling processes, but that is not the main objective.

The time that it takes to accomplish the dispersion milling can vary from as little as 4 hours to as long as 48 hours. A typical dispersion milling time as used in our laboratory is 24 hours. In some cases this may be overkill, but it is a convenient time frame to work within for a single-shift workplace, and it is found in many publications dealing with tape casting. The actual dispersion process is time dependent to some extent, since the breaking of the agglomerated particles can take considerable energy transfer if the agglomerates are of the "hard" type, that is, sintered together rather than simply attracted to one another. A secondary dispersion test can be performed to approximate the time necessary for complete dispersion of deagglomerated particles. One can use the same test described in the section on surfactants, the "dispersion study," with a slight modification. Once the proper amount of dispersant has been determined, the test can be redone using that amount of dispersant for a series of different times. Once again, a graphical representation of the data will indicate a leveling out of the settling volumes after a period of time. This is the point at which the dispersion milling procedure has reached an end. Another test would be a series of particle size analyses as a function of dispersion milling time.

A dilution of the milling solvent(s) would have to be used in any sedimentation tests, since the slurries should never be dried before testing, and a dilute solution is required for most of the tests. When the particle size analyses duplicate one another after milling for two different periods of time, the end point has been reached for the dispersion milling. As we mentioned previously, this time frame is rarely if ever longer than 24 hours. The d_{50} and particle size distribution of the dispersed powder, even after as brief a milling time as two hours, may be much different than those of the as-received powder, depending on the particle size analysis measurement's ability to detect agglomerates. As-received powders should be analyzed by SEM and by particle size analysis on an ultrasonically dispersed suspension. (See the section on powders in Chapter 2.) There are always exceptions to this rule. For example, Stetson and Gyurk, in their patent on the use of fish oil as a dispersant,[17] found that the surface area of the alumina they were dispersion milling gradually increased over a period of 120 hours but then did not increase any further up to 250 hours. No mention was made of the wear of the grinding media and mill during this process. It was determined during later experiments that considerable impurity pickup occurred during the dispersion milling process and that, potentially, some of the surface area increase could have come from this mill jar and grinding media wear.

The mill rotation speed should be at a fraction of the critical speed, which is defined by the following equation taken from the *Handbook of Ball and Pebble Mill Operation*:[18]

$N_c = 76.6 / \sqrt{D}$, where D = mill inside diameter in feet and N_c is the critical rotation speed in rpm.

Most milling operations operate in the range of 35 to 115% of the critical speed. The smaller the mill diameter, the faster the mill rotates at the critical speed. For example, the critical speed for a typical size 1 mill with a diameter of 9" (.75 foot) is 88 rpm.

For a smaller size 0 mill with a diameter of 5-7/8" (.49 foot) the critical speed is 109 rpm. Most laboratory jar rollers either have variable speed controls or are preset to fall in the 35 to 115% range. Most of our dispersion milling takes place at 50 to 80 rpm. From a practical "hands-on" perspective, if the mill is rolling at a correct speed, the sound of media impact against the jar should be audible.

The amount of material charged in the mill should follow the guidelines given by the mill manufacturers. A good rule of thumb to follow is to fill the mill with slip to about one-half to two-thirds of the total volume (with the grinding media in the mill). This will leave room for the addition of the plasticizers and binder during the next stage of slip preparation. We have processed mills in our laboratory that were filled to the brim. This is not good practice, but in some cases it becomes necessary to meet specific needs, such as tape production footage. As long as the dispersion milling procedure can occur with good action of the grinding media and as long as the binder has enough room for thorough mixing to take place, you are usually safe in slightly overcharging the mill. Overcharging, however, is not a recommended practice.

The critical point to remember with respect to dispersion milling is that it must accomplish its objective of breaking up agglomerates and separating the particles into individual entities before the next phase of slip preparation can begin; the addition and mixing of plasticizers and binder. If the objective of the dispersion milling procedure is not met, then the zipper bag theory previously mentioned can occur.

In that case, groups of particles (or agglomerates) are encased in a polymer film along with entrapped air and act as a single unit for the remainder of the tape casting process. It would be as if there were a number of these zipper bags full of agglomerated particles and entrapped air being mixed in a solvent-rich slip. If the particles are well dispersed and deagglomerated before the plasticizer and binder are added, this will not occur, and the characteristics of the slip and the tape will be optimized. An indication that deagglomeration has not occurred would be a very low slip viscosity after the binder is added. If the dispersion milling procedure has been accomplished properly, the next stage of the slip preparation process can begin: the plasticizer-binder addition.

3.3 PLASTICIZER AND BINDER MIXING

Before embarking on the procedures for adding the plasticizer(s) and binder, it is important to point out that the order in which you add the organics to the mill is almost as critical as the materials themselves. The dispersant, which is dissolved in a solvent or solvents, is added to the mill first, followed by the powder to be dispersion milled. After the dispersion milling is completed, the plasticizer or plasticizers are added to the slip, followed by the binder, either in powder or dissolved form. There are formulations in the literature and procedures described by binder manufacturers that do not follow this step-by-step procedure; indeed they make the steps impossible to follow, since the binder solutions are sold as a single component. We feel very strongly about the procedures outlined in this book, because years of experience have shown us that tapes with

Zipper Bag Theory

There are two main groupings of dispersants: steric hindrance and ionic repulsion. Steric hindrance describes the attachment of a molecule or molecules to a particle surface and building a physical barrier around the particle, which extends past the primary minimum (as defined by DLVO theory). In a tape casting slip containing a dissolved polymer binder, the binder is a powerful steric hindrance dispersant. The dissolved polymer molecule, however, is typically much larger than the particle itself and can attach to multiple particles or sites on a single particle. If the powder is not first deflocculated and/or deagglomerated in the solvent vehicle, or if the binder is added prior to deflocculation in the solvent vehicle, then the binder can attach to multiple particles of an agglomerate prior to deagglomeration and bind the agglomerate together.

This is like putting a clump of particles inside a bag, sealing it shut, and then trying to get the particles apart. While it is possible to break the agglomerate within the bag, all pieces of that agglomerate still remain within the bag until the bag is broken.

Adding the binder after deflocculation and deagglomeration is, using the same metaphor, putting each particle into a separate bag. It is a well-known fact that a well-dispersed and deagglomerated slip settles to a denser packed bed than does a slip that is only partially flocced or agglomerated.[19] The authors and others[20,21] have shown that higher densities are achieved with identical slip recipes when the binder is added to a deflocculated and deagglomerated suspension.

better properties result if they are followed. One can cast tapes using one-component binder systems, but the optimum tape properties are never reached. If defect-free tapes are required and if tapes of 10 microns or less are the goal, then two-stage milling and mixing procedures are essential.[22,23,24,25,26]

After the dispersion milling process is complete—that is, after 24 or more hours of milling—the mill is opened and the plasticizer or plasticizers are poured onto the top of the slip. Most plasticizers are liquids and quite fluid, and therefore this is not a complicated procedure. The plasticizer or plasticizers form a thin film on the surface of the slip in the mill. The weighed quantity of binder, either in powder form or dissolved in one or both of the solvents, is then added to the layer of plasticizer(s). Usually the binder is more soluble in the plasticizer(s) than it is in the solvent(s), so it is good practice to introduce the binder to the plasticizer first. Remember from Chapter 2 that the whole point of the Type I plasticizer is to react with (dissolve) the binder. The plasticizer wets the binder if it is in powder form and aids in the dissolution process. It is also good practice to make sure that the binder, if in powder form, is stirred into the slip mixture before the mill is sealed and returned to the rollers. We have found large clumps of undissolved binder after 12 hours or more of mixing if we didn't follow this simple procedure.

Once the plasticizer and binder are in the mill and the mixture is stirred to make sure that the binder is not in large clumps, the mill lid is replaced and the second stage of mixing is started. The mill is placed on the rollers and is rotated at the same speed used for the dispersion milling procedure. When "mixing" in the binder is not possible—for example, in a 300-gallon mill—it is sometimes beneficial to add gradual amounts of the binder to avoid "clumping" of the binder.

Usually about 12 hours are required for the complete dissolution of the binder and reaction of the binder with the Type I plasticizer (see the definition of a Type I plasticizer in Chapter 2). If the binder is pre-dissolved in a solvent mixture and the Type I plasticizer is added during the dissolution, this time can be reduced to about four hours, since the binder is both pre-dissolved and pre-reacted in this case. One does not always have the luxury of being able to pre-dissolve the binder, since at times the amount of solvent in the slip is low and all of it is required during the dispersion milling process. The time used for the binder dissolution should be determined by experiment, since there are many formulations in which the binder acts as a secondary dispersant and actually replaces some of the dispersant on the surface of the particles. This is particularly true of binders in the polyvinyl butyral class. In recent years, several papers have been written about the interaction of the organics in tape casting formulations,[27,28,29] and we have reviewed some of this work in Section 2.6. In previous work we have actually used polyvinyl butyral as a dispersant for aluminum oxide.[30] There are some very high surface area powders that require the addition of a small portion of the binder with the dispersant during the dispersion milling procedure to keep the viscosity low, even with a relatively high solids content. This should be done with care to minimize the zipper bag effect.

Some slips become less viscous with time of milling and others become more viscous. The ideal situation is to mix the plasticizers and binder into the well-dispersed slip until they are completely dissolved, and then stop the mixing procedure. If one uses the same duration for this procedure from batch to batch, the viscosity and other slip properties will remain fairly constant. Usually the slip will become more viscous with the addition of a long-chain polymeric binder. The only time that this does not occur is when the binder is a more powerful dispersant and replaces or overrides the dispersant of choice. This is also a good reason for using the slip in the tape casting process as soon as possible after the binder is mixed into the slip. Rheological properties of the slip can and do change with time.

In a study conducted at Western Electric in the 1970s the effect of different times for each of the stages of slip preparation on the final fired density of the aluminum oxide substrate material was determined.[31] Figure 3.1 shows the results of that study, where fired density is plotted as a function of both total time in the mill and the split of the time between the dispersion mill and the binder/plasticizer mixing. The first number at each data point is the dispersion milling time, and the second number is the binder/plasticizer mixing time. The significant results from this work indicated that low fired densities resulted from insufficient dispersion milling time or from excessive mixing time in the presence of the binder/plasticizer. Insufficient dispersion milling time is quite easy to explain: The agglomerates have not been broken completely and the dispersion is incomplete (zipper bag theory). Both of these effects would cause low unfired densities and therefore low fired densities. The low densities resulting from too long a mixing procedure in the presence of the

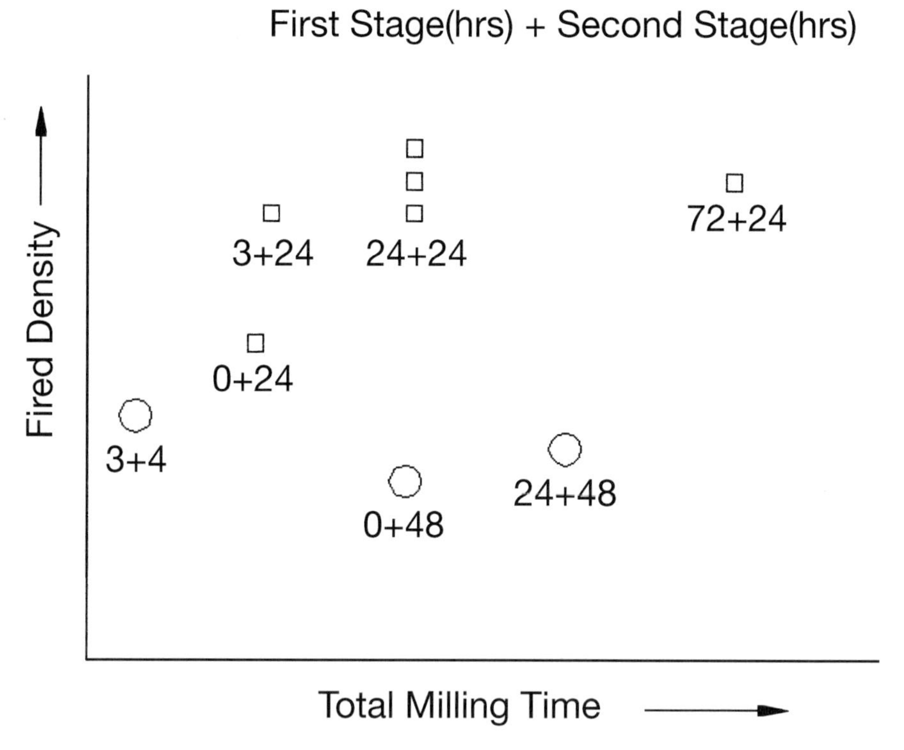

Fig. 3.1 The effect of different times for each of the stages of slip preparation on the final fired density of the aluminum oxide substrate material. Fired density is plotted as a function of both total time in the mill and the split of the time between the dispersion mill and the binder/plasticizer mixing.

binder/plasticizer is harder to explain. One possible explanation is the substitution of the dispersant over a period of time by the long-chain binder (polyvinyl butyral), which could have caused a lower packed density in the unfired tape and therefore a lower fired density. Another effect that was observed was the low fired density when all of the ingredients were milled together without a dispersion milling procedure. (Zero time for the first step.) This supports our contention that a two-stage procedure is the proper course of action.

Once the plasticizer(s) and binder(s) have been mixed into the slip formulation, it is ready for precasting conditioning and characterization.

3.4 SLIP DE-AIRING

The slip at this point has all the ingredients in an intimate mixture and is ready for the casting operation. The next step in that process is to de-air the slip to remove any air that may have been entrained during the milling and mixing process. Air bubbles cause defects in the tape-cast product. Pinholes (small holes left by air bubbles) are the most common defect, and these can lead to "crows foot" cracking (cracks that radiate from the pinhole) upon drying. Other defects attributable to air bubbles, especially in thin tapes, are elongated streaks or thin spots in the tape in the casting direction. It is therefore essential that entrained air be removed before the casting process. Several techniques are used for the de-airing process.

By far the most common technique for de-airing a slip for tape casting is to use a partial vacuum accompanied by gentle stirring or agitation. Agitation tends to lower the viscosity in a pseudoplastic slip, and therefore it makes the air removal easier. (Pseudoplasticity will be covered in detail in Section 4.2.) This is accomplished in the laboratory in a vacuum desiccator using a vacuum in the range 635 to 700 mm of Hg. This is a low vacuum and can be generated with a rough pump or an air aspirator (venturi pump) based upon the Bernoulli principle. Too high a vacuum will tend to remove a large volume of solvent along with the air bubbles. In some laboratories, the actual volume of solvent removed is captured in a cold trap and is recorded in order to keep the solids loading constant from batch to batch. This is not a recommended procedure, although there is certainly no harm in it. If the vacuum is kept low and the time is kept constant for a given volume of slip, the results will be consistent from batch to batch and run to run.

Depending on the volume of slip to be de-aired, the time required will range from eight minutes to more than one hour. As a general rule of thumb, four liters of slip will usually be de-aired in about five to eight minutes under these conditions. Careful visual observation of the bubbling action of the slip surface during the de-airing process can usually be used as a gauge of the completion of the procedure. The bubbling is usually very vigorous during the initial phase, when there is a lot of entrapped air present. The bubbling gradually decreases to a point where most of the air has been removed. Evaporated solvent bubbles can form if the process is continued for too long. It is almost impossible to determine when this end point has been reached; therefore the use of a constant time for de-airing is recommended. The slip viscosity also comes into consideration during de-airing. High-viscosity slips are much more difficult to de-air than low viscosity slips, and they require more time and agitation during the process.

In a production-scale procedure, the de-airing is done in a large tank equipped with an air-driven stirrer. The rotation of the stirring blade is very slow (usually 10 rpm or less) to prevent any cavitation effects and yet prevent any settling during de-airing and casting. The size of the de-airing tank is usually determined by the quantity of tape to be cast, since the same tank is used as a pressurized feed system for the tape casting operation. We have worked with tanks as large as standard 200-liter containers (55 gallon drums), which fed several tape casting lines simultaneously. Most small-scale production lines use containers that hold 10 to 20 liters of slip. These vessels have wheels so that they can be brought to the casting line, and they are self-contained units that have portholes and lights for observing the bubbling action during de-airing. Some have heating bands to maintain a constant temperature during the casting operation.

Other techniques have been used to de-air slips before tape casting. One of these techniques involves slow rotation on a set of rollers for 24 hours or more. The usual rotation speed in this case is 10 rpm or less. This is the technique used to keep screen printing inks in suspension for long periods of time and also to remove any air bubbles from very viscous slurries. Another technique used for slips that cannot tolerate negative pressure or that tend to create foam during vacuum de-airing is simply a slow stirring procedure in a closed container. Once again, this is usually done for 24 hours or longer. The vacuum de-airing process described previously is the most effective procedure, and it is used by most large-scale tape casting houses in the world today.

3.5 SLIP CHARACTERIZATION

At this point in the manufacturing sequence, a series of characterization procedures are usually inserted to check the slip quality and uniformity. They are standard in-process quality control checks to make sure that the slip is the same as the batches that preceded it into manufacture. The characteristics monitored are:

- viscosity
- specific gravity
- particle size distribution

The first two, viscosity and specific gravity, are determined on every batch of slip brought to the casting line. Particle size distribution, on the other hand, is usually only checked if there has been a change in the lot of inorganic material or materials in the batch.

Viscosity can be checked using any number of instruments, ranging from very sophisticated cone and plate techniques to production-line quality control rotating spindle techniques. Standardized procedures, using the same instrument and the same technique, are recommended to produce meaningful data. In our laboratory we have standardized on the use of a Brookfield viscometer with an RV4 spindle rotating at speeds of 10 to 100 rpm, depending on the slip. Most tape casting slips fall in the range of 500 to 6000 mPa·s (cP) when tested using an RV4 spindle at 20 rpm. The 20-rpm rotation speed is used by many laboratories, since popular belief has it that this approximates the shear rate under the doctor blade during casting and allows batch-to-batch comparisons for quality control purposes. We apologize for helping to spread this myth! The viscosity, doctor blade gap, doctor blade shape, and speed of casting all come into play when the shear under the blade is approximated. In many cases it is quite different from that generated by an RV4 spindle at 20 rpm. We feel that the "real" reason for the use of an RV4 spindle at 20 rpm is the fact that an easy multiplier of 100 yields the viscosity number in mPa·s (cP) when these settings are used. For quality control purposes, the same test should be used for every batch of slip. If the batches are experimental in nature, then more sophisticated techniques should be used to determine the rheological nature of the slip, that is whether it is shear-thinning, pseudoplastic, thixotropic, and so on. Most tape casting slips are pseudoplastic and exhibit a decrease in viscosity with an increase in shear rate. This behavior is shown in Figure 3.2. As can be seen, the viscosity tends to decrease as a function of increasing shear rate.

Specific gravity is a very easy test to perform, and it is recommended for quality control for every batch of slip brought to the tape casting line. Usually, 100 cc of slip is poured into a graduated cylinder, and the weight of the slip is determined. The weight of the slip is divided by the volume (100 cc) to yield the specific gravity in grams per cubic centimeter. Batch-to-batch variation should not vary more than ± 0.02 g/cm^3. Since most tape casting processes involve the use of volatile organic solvents, the weighing procedure must be done rapidly to prevent the loss of solvent during the measurement.

Particle size distribution analysis can be performed on every batch of slip, but most tape casting operations only use this test when new or different lots of starting inorganic powders are introduced into the mixture. The procedure used for the determination of particle size distribution for tape casting slips is different from standard techniques. We feel that it is critical to perform this analysis on the slip as it is to be used in the process; in other words, the slip should never be dried. As a matter of fact, a sample of the slip can be taken from the container after de-airing for use in this analysis. The sample is diluted with a suitable solvent and then a standard sedimentation particle size analysis is performed.

As a representative example we used a slip prepared with a high-surface-area aluminum oxide with trichloroethylene and ethyl alcohol as the solvents, menhaden fish oil as the dispersant, and polyvinyl butyral, polyethylene glycol, and octyl phthalate as the plasticizers. This batch was prepared using the two-stage milling and mixing procedure described previously, and then a sample was taken for analysis. The sample, 7.56 grams of slip, was diluted in 40 cc of a

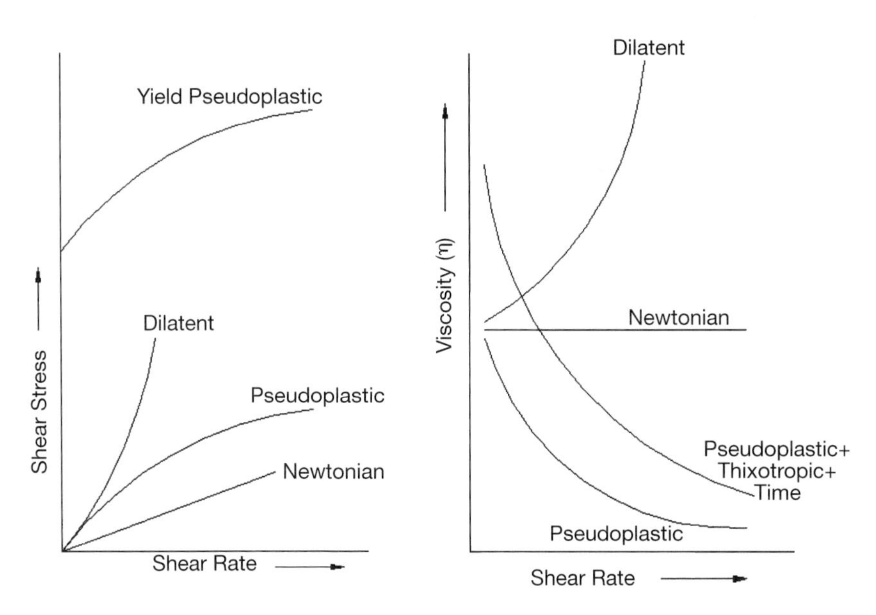

Fig. 3.2 Most tape casting slips are pseudoplastic and exhibit a decrease in viscosity with an increase in shear rate. Viscosity tends to decrease as a function of increasing shear rate.

50 vol% solution of benzene in ethyl alcohol, and this diluted sample was analyzed using a standard sedimentation technique.[32] The results for this material indicated a median particle size of 0.4 micron, equivalent spherical diameter. Several tests were performed using different samples, and the range of precision was determined to be ± 0.04 micron using this technique. This kind of test can be used for most tape casting systems if the proper diluent solvent(s) can be found. The laboratory technician can usually find miscible solvents to accomplish this. The sedimentation technique yields a particle size distribution curve that can be compared with previous lots of starting powder.

The optimized slip is now ready to be moved to the casting equipment for forming the tape-cast product. The tape casting process is discussed in Chapter 4.

CHAPTER 4

The Tape Casting Process

In this chapter we will concentrate on the process of forming tapes. We will discuss the equipment used in doctor-blade tape casting as well as competing processes such as roller coating, micro-gravure coating, and slot die coating. The slip delivery and filtration processes will be discussed along with their effect on the slip rheology just prior to the casting process. The types of carrier or substrate materials used for continuous processing of tapes will be reviewed, as will suggestions about their utility in various solvent systems and for various powders. Tape casting machines will be described, and the need and appropriate use for the various options will be reviewed while keeping in mind the physics of the tape casting process and of the drying process prior to tape removal. A detailed section about the physics and procedures involved in the flat-bottomed doctor-blade tape casting process is included, since that is the procedure upon which we have concentrated our efforts. That section is followed by a detailed analysis of the drying process and the changes that take place in a cast film as a function of time, temperature, air flow, and degree of air saturation with solvent vapor. The final section in this process-oriented chapter deals with the removal of the tape-cast product from the tape casting machine. First we will discuss the last precasting procedure used in most tape casting operations: slip delivery and filtration.

4.1 EQUIPMENT

4.1.1 Slip Delivery and Filtration

Slip is usually delivered to the tape casting machine, and specifically to the doctor blade reservoir, either through pipes or hoses. The selection of the type of material used depends on several factors: the type of solvent(s) being used and how it (they) react with the material of choice, the ease of cleaning the system during a shutdown of the process (i.e., is it easier to discard the tubing or to go through an elaborate cleaning procedure?), and the ability of the tubing to withstand the pressure generated by the delivery system. We have seen delivery systems that were "hard-piped" using standard galvanized steel plumbing pipes, which were then connected to Tygon™ tubing, which fed the doctor blade reservoirs. As many as six tape casting lines were operated from one pressurized master feed tank using such a system. Other processes use flexible polymer tubing directly from the pressurized de-airing chamber to the filter and then to the doctor blade reservoir. Still others use some sort of peristaltic "finger" pump to deliver the slip through the flexible polymer tubing.

The majority of the systems in production today use some sort of polymer tubing to deliver the slip from a pressurized holding tank. In these cases, the tubing is discarded at the end of a casting run and is replaced with new tubing at the beginning of a new run. Several types of polymer tubing can be used, as long as the material is resistant to the solvents being pumped through it. There are different grades of standard Tygon tubing that are resistant to solvents such as xylenes, alcohol, and methyl ethyl ketone. One should check the reactivity of the tubing selected, even when the literature deems it resistant to a solvent. A simple but effective test is to soak the tubing in the solvent for a period of time, usually several days, and observe the swelling characteristics. This is especially true when pumping the slip through the tubing using a peristaltic pump. We have seen tubing weaken and actually "blow up" due to the repetitive action of a finger pump. With a peristaltic pump, the tubing must be flexible enough for the pump to be effective but resistant to the combined pumping and solvation effects. For pressurized feed systems, the tubing must withstand the pressure that can be generated, especially on the feed side of a filtration system (filtration will be discussed in later paragraphs).

The use of "hard" piping up to the filter precludes the use of a finger pump but eliminates the pressure buildup problem. It also introduces a complicated cleanup procedure that must be performed immediately after the casting is completed. Usually, solvent is flushed through the system as a cleaning method. Although the solvent can be reused for cleaning, it usually is not, and this introduces another source of waste solvent that must be either reclaimed or disposed of in a proper manner. For most of the slips processed in our laboratory over the past 15 years, the use of standard Tygon or methyl ethyl ketone–resistant Tygon proved adequate over a short period of time.

Pressurized slip feeding is usually accomplished using very low pressures. In experiments performed in our laboratory, we found that the pressure to move the slip rarely exceeded 0.03 MPa (5 psi). This applied to slips with viscosities ranging from 500 to 4000 mPa·s (cP) and with specific gravities as high as 6 g/cc. In all of these experiments the tubing utilized was a 9.52 mm (3/8 in.) ID Tygon material. The pressure (or in the case of the peristaltic pump, the pumping rate) is adjusted to keep the flow of the slip constant to the doctor blade reservoir. The actual flow is usually controlled using a sensor and valve system, which will be described in detail later.

As mentioned in the slip preparation section, it is very important to maintain a constant temperature in the conditioning and delivery tank during the tape casting process. The use of a temperature slightly above room temperature is sometimes recommended in order to lower the viscosity to facilitate air removal during the de-airing procedure. The slip conditioning systems used in manufacturing are sometimes equipped with heating blankets or bands to provide this constant heat source. It is also recommended that the slip in the conditioning tank be stored under the same temperature conditions as will be encountered in the casting room. Some manufacturers actually keep the tanks of slip, which are under constant agitation, in the same room as the casting machines. The use of heating bands or other equipment to raise the temperature of the slip above room temperature during a cast has been suggested by the authors in order to maintain a constant-viscosity slip in the reservoir. A schematic drawing of a typical de-airing/pressurized feed tank is shown in Figure 4.1. This schematic has all of the essential features described in the previous sections on de-airing and slip delivery.

After the slip is either pressure fed or pumped from the holding (de-airing) tank, it usually, but not always, passes through a filter of some sort before flowing into or behind the doctor blade assembly. The first description of such a filtration system was published by one of the authors in 1976.[1] In that paper, the basic reason for the use of a filter was reviewed: the removal of "Imperfectly dispersed material in slip. . . . a potential cause of surface defects in slip cast ceramic bodies." The incorporation of a filtration process into the tape casting process accomplishes even more than is spelled out in that paper. The removal of ceramic agglomerates is only one of many functions of the filter. It also removes agglomerated (and undissolved) binder, pieces of the mill lining or media that have chipped

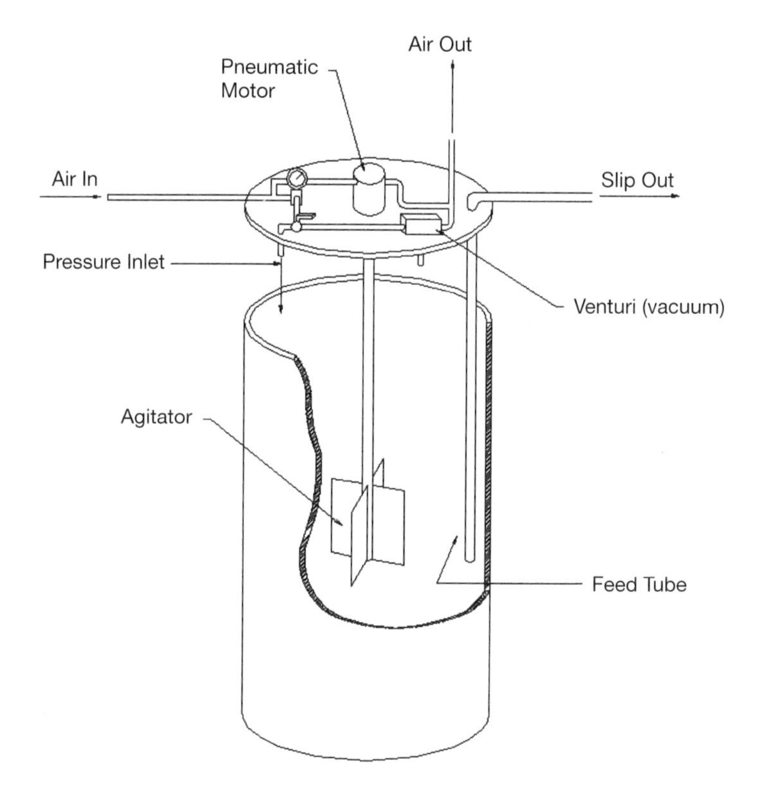

Fig. 4.1 Schematic of typical de-airing/pressurized feed tank.

off the surface, and any air bubbles that remain after de-airing. Some bubbles are broken down into smaller bubbles by the filter medium, but these are usually so small that they do not cause any defect problems in the dried and sintered ceramic or metal part.

A wide variety of filters have been used for tape casting. One of the most common is the one described in the paper cited previously.[1] Figure 4.2 is a schematic drawing of this device. The filter frame is made from aluminum discs with shallow conical recesses on one

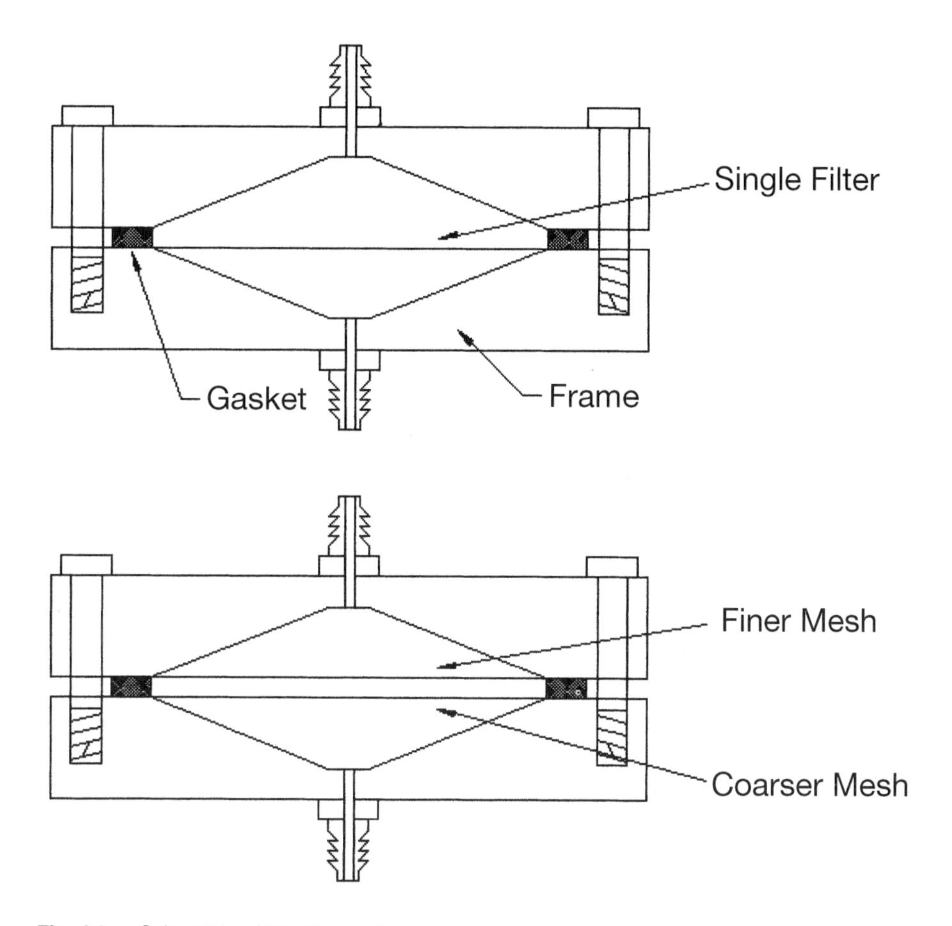

Fig. 4.2 Schematic of filtration system.

side. The filter material is clamped between the aluminum discs using a Viton™ or Solvaflex™ gasket, depending upon the resistance to chemical attack needed for the solvent system being used. Viton is used for solvents such as xylenes, ethyl alcohol, and toluene, and Solvaflex is used for methyl ethyl ketone and other ketone-type solvents. The filters used are standard polyester monofilament screening fabrics. The mesh openings are available in just about any size down to 5 μm. The use of such a fine filter is not usually recommended or needed for most processes. In the Western Electric process for very smooth, low-defect thin-film substrates, a dual filter was used for the alumina slip: a 37-μm filter and a 10-μm filter in series. The use of the dual filter eliminated the need to constantly change the filter, since the coarser screen prevented the rapid blockage of the 10-μm screen.

Other filters used on slip delivery systems range from simple gasoline fuel filters from an automotive supply house to water filter cartridges. Both of these types of filters can be purchased with different grades of filter papers. The gasoline filters are throwaway items, and the water filter cartridges can be thrown away and replaced with new cartridges. The criteria for selecting the filter and the filtration medium are its reactivity (or lack thereof) with the solvent system of choice and its durability in the process.

In the design of the filtration station, one must consider the duration of the casting process. Filters will plug up over a period of time. In a laboratory setup, the filter can either be in-line with the casting process or it can be "off-line," that is, filtration can be done before the slip is brought to the tape casting line. This is the usual practice for short experimental casts for which the slip is added to the reservoir manually. We have also used a very simple system for filtering very small volumes of slip. The slip is poured into an HDPE (high density polyethylene) jar, a hole is cut into the jar lid, and a piece of filter cloth of the proper mesh size is held in place by the screwed-on lid. The jar is then squeezed and the slip is collected in a beaker. We have also used a very small HDPE jar with a nozzle top with a piece of filter cloth held in place by the screw top. In this case the slip is squeezed directly into the doctor blade reservoir.

In a production process, a series-parallel setup is highly recommended. In that setup, two filters in parallel are attached on one side through a Y connection to the feed system and on the other side through a Y connection to the doctor blade reservoir. There are shutoff valves on each side of this parallel system so that the slip flow to the doctor blade can continue uninterrupted through one filter while the filter on the other side is being changed. This type of system is currently being used on production lines throughout the world.

The pumping and filtration procedure serves another purpose in addition to filtration. The movement of the slip through a tube and then through a filter also keeps the slip fluid. This is known as shear-thinning; the viscosity actually decreases as a function of the shear rate. The physics of this phenomenon is described in Section 4.2. Slip that is not kept in motion tends to form "short-range order"—that is, it tends to thicken. This will also be described in more detail later in this chapter. The pumping and filtration process keeps the slip fluid and prevents the formation of this short-range order effect.

During continuous casting, such as that used in a production mode, the slip is delivered from the filter to the doctor blade reservoir through a pipe or tube that has some sort of valving mechanism in series. This valve controls the flow of slip to the reservoir and helps to maintain a constant "head," or in some cases puddle, of slip behind the doctor blade. The control of the level of slip in a reservoir-type doctor blade assembly is one of the variables that combine to regulate the wet thickness of the tape being cast. This subject will be discussed in detail in Section 4.2 on the physics of tape casting.

One problem only occurs with multiple feed tubes from a manifold system. This type of feed system is employed when the doctor blade width is 400 mm (about 16 in.) or more. The use of a manifold feed system is necessary to keep the slip level in the reservoir at a uniform depth along the width of the doctor blade assembly. The problem that arises is the blending of the slip streams that flow down the rear wall of the reservoir. Streaks are generated on the surface of the cast tape where the slip streams come together during the feeding operation. These streaks are caused by the higher solvent concentration at the slip:air interface for each of the streams, and this solvent-rich interface is carried through to the cast tape, even after it passes under the doctor blade. The defect manifests itself as thin lines or streaks.

Many different techniques have been used to either minimize or eliminate this effect. One technique involves the use of a mixing mechanism in the slip reservoir. Another feeds the slip streams through a screen mesh. (This usually only breaks the flow into much smaller streams and produces streaks that are much closer together and harder to see.) The best technique is to use a feed system that does not introduce the problem in the first place. The slip is fed to the reservoir under the surface, not on the top surface. This has been accomplished by the use of a reservoir system like that shown in Figure 4.3. This reservoir also uses a "weir" feed system, in which the slip must flow from the first chamber over a retaining wall into the main feed chamber of the doctor blade assembly. The flow down the retaining wall is similar to a waterfall and provides a uniform and

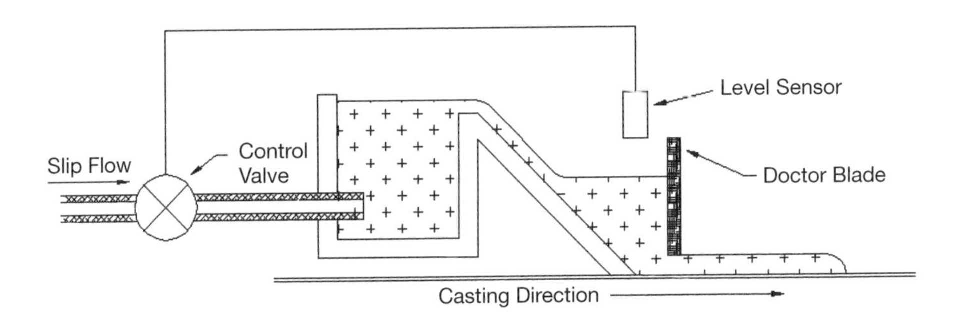

Fig. 4.3 Schematic of weir-type doctor blade assembly.

constant source of slip to the main reservoir behind the doctor blade. We have seen weir-type systems with as many as three chambers with a tortuous path for the slip to pass from the feed tubes to the doctor blade. The main purpose (which is to eliminate the streaks) is accomplished, however, by feeding the slip to the reservoir from beneath the slip surface.

There are many different combinations of sensors and control valves on production machines. The sensor is usually mounted above the slip in the reservoir behind the doctor blade. In the case of "puddle" casting, where the slip is pumped onto the carrier behind the doctor blade, there is no reservoir, and the sensor detects the spread of the slip backward from the doctor blade itself. The most common sensors are based upon capacitance or inductance and are known as proximity sensors. These sensors detect the surface of the slip and generate an electrical impulse when the slip is within range. The electrical impulse in turn controls either an on-off valve or a proportional valve to control the flow of slip to the reservoir. When the slip falls below the range of the sensor, the valve automatically opens to allow more slip to flow into the reservoir until the electrical impulse begins the controlling cycle once more.

There are other types of sensors in use today. There are fiber-optic sensors that can detect reflections of light from the slip surface, and there are air-jet sensors that can detect changes in the air pressure emitted as a function of distance from the slip surface. The main concern in sensor selection is chemical resistance and reliability, since the detection mechanism must be fool proof, so that the casting remains constant over a long period of time. We have seen a system based on the low-pressure air jet detection sensor. This sensor was coupled with a pinch-valve-controlled pressurized polymer tubing feed system. On more than one occasion, when left alone for a few moments, there was either slip flowing over the reservoir sides or the reservoir had gone completely dry. This was a very unreliable control system, but it was considered very "safe" since there was no electrical signal in the vicinity of the casting head. The signal generated by the low-pressure air flow was very weak, and unless amplifiers were incorporated it would not work well over time. For this reason, most systems today use capacitance or light reflection.

The positioning of the sensor over the slip surface must be exact from cast to cast and must not change during a casting run. This is the only way to maintain a constant surface level, since the sensor has a distinct range of operation for maximum detection efficiency. Most sensors are secured by a locking nut or other mechanism. The best way to control the slip flow to the doctor blade is with the use of a continuously variable proportional valve. This valve can meter the flow of slip to the reservoir and maintain a very constant head behind the doctor blade.

Another technique for controlling the reservoir slip height behind the doctor blade is to use a dual doctor blade system. This dual doctor blade is shown in Figure 4.4. The dual doctor blade has been described in the literature[2] as a means to control the hydrodynamic forces behind the casting blade (the blade to the left on Figure 4.4). Because of the hydrodynamic forces, which will be described in much more detail in Section 4.2, it is necessary to maintain a constant height in the casting pool at the right side of the casting blade in order to maintain a constant height of cast tape to the left of the blade. In order to

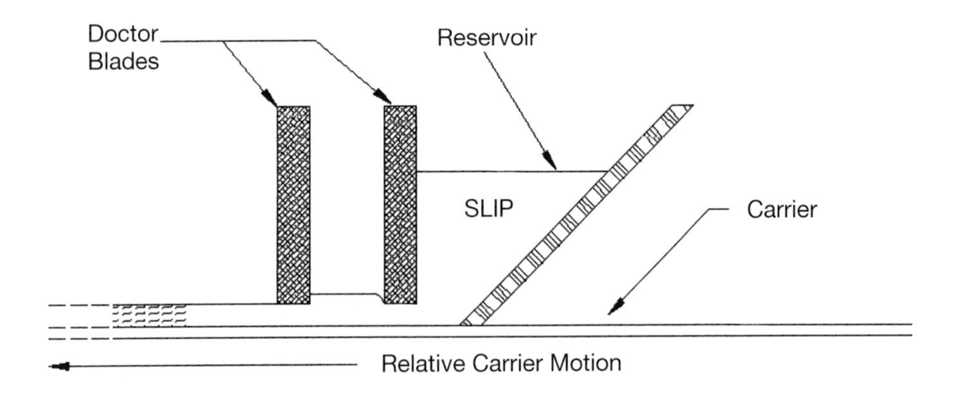

Fig. 4.4 Dual doctor blade assembly.

avoid "welling" (described in the next section), it is also helpful to minimize the height differential between the casting pool and the cast tape. The simple two-blade design satisfies both of these conditions. By adjusting the relative heights of the two blades and the speed of the casting carrier surface, one can maintain a constant low-level casting pool height behind the casting blade. In such a dual doctor blade system, it is not as critical to control the reservoir depth behind the blade to the right (which we prefer to call the metering blade, since it meters the amount of slip behind the casting blade).

Of course for some other systems there is no free-standing reservoir that feeds the doctor blade or other coating device. In these systems, there is usually a pressurized manifold chamber that feeds the coating head. The amount of flow to the coater in these cases is cotrolled by the pressure itself. Some of these coating systems will be described in the next section.

4.1.2 Doctor Blades and Coating Heads

There are as many coating systems in use today as there are companies that manufacture them. The standard doctor blade has been in existence from well before the time Glenn Howatt first used it for ceramic casting. It was originally used for paper coating systems in large pieces of equipment that resemble tape casting machines. About the only difference from doctor blade to doctor blade is the blade shape where it makes contact with the slip being spread onto the substrate or carrier surface. The shape of doctor blades and the evolution of shapes is described later in this chapter. At this point we will only mention that there are thin, flat-bottomed blades, that is, those less than 12.7 mm (1/2 in.) thick, thick flat-bottomed doctor blades, (those 12.7 mm and thicker), round-bottomed doctor blades, "comma"-shaped doctor blades, beveled-bottom doctor blades with the bevel on one edge, beveled-bottom doctor blades that form a knife edge, and combinations of these. These are schematically depicted in Figure 4.5. The physics of flow under the doctor blade is the subject of Section 4.2.

Another difference from doctor blade to doctor blade is the material from which the blade is manufactured. Most commonly, blades are made of stainless steel or hardened carbon steel. Some manufacturers of large amounts of abrasive tapes use blades made from tungsten carbide or aluminum oxide. No matter which material is used, it is important that the casting surface be ground flat and smooth. (That is for a flat-bottomed blade). If the entire doctor blade assembly is designed to "ride" on the carrier surface, then weight becomes an important issue. The reservoir section and the frame that holds the doctor blade are usually made from aluminum sheet. In some cases we have used Teflon™ sheet for the side rails of the assembly—that is, the parts that ride on the carrier—to reduce friction between the casting box and the carrier surface. The Teflon box is also very easy to clean after a casting is completed. The design of doctor blade assemblies should be such that they can be easily disassembled for

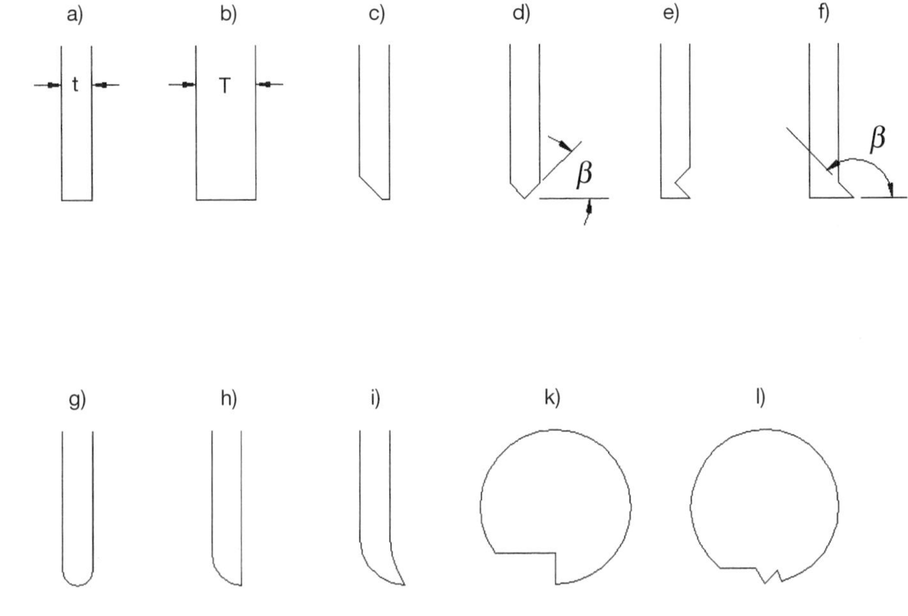

Fig. 4.5 Doctor blade and bar shapes.

cleaning. At the very least, the blade should be readily removable from the rest of the casting box.

In general, doctor blades are used for casting thicker tapes, that is, those that are 25 microns or thicker in the dried state. All ceramic substrate casting machines use a doctor blade of some sort. For these applications, the "green" tape thickness is usually on the order of 250 to 1000 μm (0.010 to 0.040 in.). Some of the machines produced in Japan[3] use the "comma" coating head for these thick tapes.

In some cases, the position of the doctor blade over the carrier during the casting process determines the shape of the bottom of the blade itself. On some tape casting machines the doctor blade rides on the carrier on a flat surface such as a granite block. Other doctor blade assemblies are mounted on either side of the carrier in an "outrigger" design. Much of the weight of the doctor blade is taken off the carrier surface in this configuration, but it is much more difficult to set up the blade gap over the carrier surface. On other machines the doctor blade rides on the carrier, which rides on a curved surface. In other words, the casting takes place on some sort of a break roller. Most of the "comma" coating blades operate on a break roller system. The use of a break roll is intended to keep the carrier flat during the coating process. There are doctor blade fixtures that conform to the shape of the break roller, while the doctor blade itself is flat bottomed. Many of these blades are manufactured with a fixed gap as shown in Figure 4.6. These blades, which are designed according to actual wet-to-dry tape thickness measurements, are used mainly for continuous thin tape manufacture. The blade gap is actually a precision-machined three-sided rectangular opening. Most of these blades have two or more different machined gaps so that a variety of tape thicknesses can be produced from one doctor blade assembly.

In some manufacturing settings and in most laboratory operations, the doctor blade gap is not fixed and can be adjusted using micrometers. This type of doctor blade assembly is shown in Figure 4.7. The doctor blade gap is usually set by using feeler gauge metal strips on a flat surface, such as a granite block. The final adjustment is then made with the micrometers. Some practitioners prefer to close the blade gap to zero on the flat surface and then back off to the desired setting using the micrometers. We have always felt that it is easier and more accurate to set the blade using feeler gauges.

As a general rule of thumb, the doctor blade coater or knife coater and the "comma coater" or other round-bottomed coating tools are used for the manufacture of tape thicknesses from 50 to > 1000 μm (0.002 to > 0.040 in.). When thinner coatings are required, the industry has turned to lip coaters, microgravure coaters, and slot-die coaters. The lip coater is used to make multilayered ceramic capacitor tapes for thicknesses of 50 μm (0.002 in) or less. For tape thicknesses 25 μm (0.001 in.) and less, the microgravure and slot-die techniques have been used very effectively.

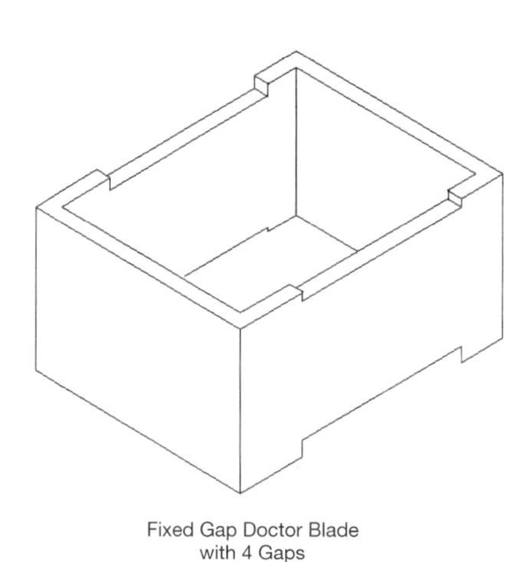

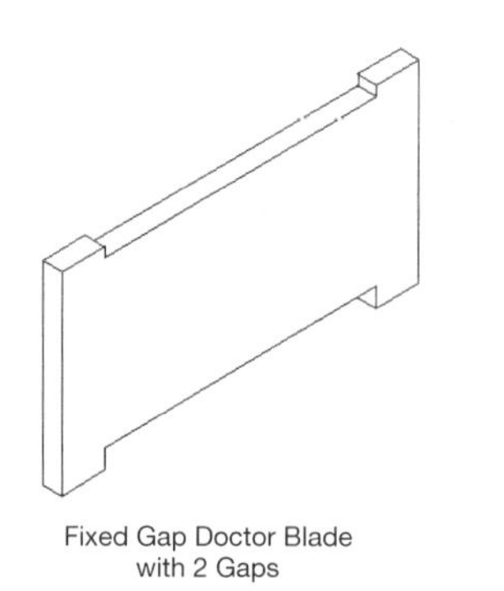

Fig. 4.6 Fixed gap doctor blade boxes.

A schematic of the basic principle of a lip coating system is shown in Figure 4.8. The slip is delivered to the carrier on the bottom side of a precision-ground roller that rotates in the casting direction. The pressurized flow of slip, which is controlled by a metering pump, gives final control of the thickness of the coating on the carrier. The "lip" is in reality a form of doctor blade, since it forms a gap between the carrier and its surface that partially determines the wet thickness of the slip on the carrier surface. Pumping speed (slip pressure), carrier speed, and the lip gap are the variables that can be controlled during a casting operation and that control the final wet thickness of the tapes formed.

A very simplistic schematic of a microgravure coating head is shown in Figure 4.9. Microgravure coating is used extensively in the printing and coatings industries.[4] It is not used very much in the coating of ceramics, but it is being experimented with in laboratories for very thin coatings (those of < 5 μm or 0.0002 in.). In the microgravure process, the slip is pumped into a reservoir into which the small-diameter gravure roller rotates. The gravure roller is engraved with a pattern that provides a specific coating volume. There is a "doctoring" blade that is very flexible that pre-meters the quantity of slip on the surface of the gravure roller. The gravure roller rotates in the reverse direction to the web being coated. This process is called reverse "kiss coating." The coating is being applied in a shearing manner rather than by printing each engraved cell, as in a direct gravure process. The microgravure terminology refers to the small size of the gravure roll, which is in the range of 20 to 50 mm (0.8 to 2 in.) in diameter. The ratio of web speed to the circumferential speed of the gravure roller is critical in establishing a uniform coating weight

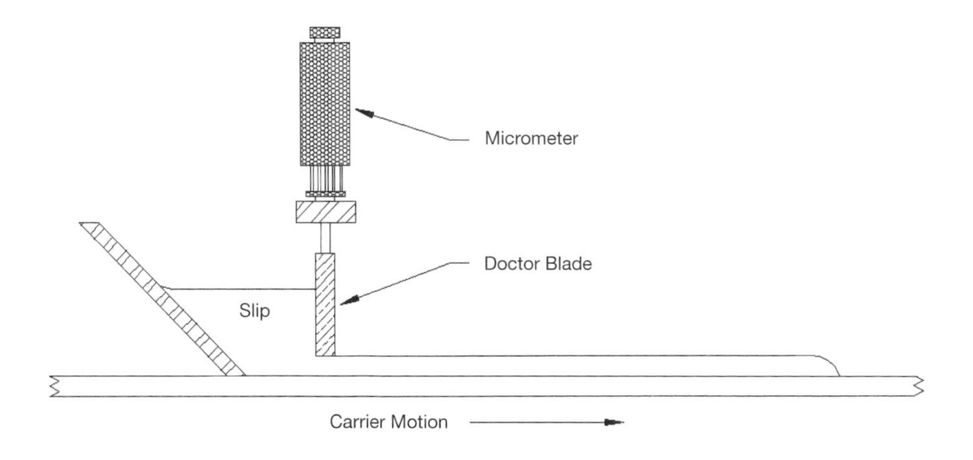

Fig. 4.7 Standard doctor blade assembly with adjustable blade.

(thickness) on the carrier. The main advantage of microgravure coating is that the process can put down thinner coatings on thinner web carriers than any other system. The reverse or shearing action of the gravure roller produces very smooth coatings. For these reasons, this technology is being investigated in laboratories around the world in an attempt to produce tapes of less than 5 μm (< 0.0002 in.).

A basic schematic of the slot-die coating head is shown in Figure 4.10. The slot-die coating technology is being used in the manufacture of ceramic, metal, and polymer coatings.[4] The major difference between this technique and the T-die method is the point of entry of the slurry to the coating head. For the slot die, the slip is introduced from the side of the die, and the uniformity of the pressure across the slot is controlled by the geometry of the interior chamber that feeds the slot. The slot die is actually an extremely accurate extrusion die

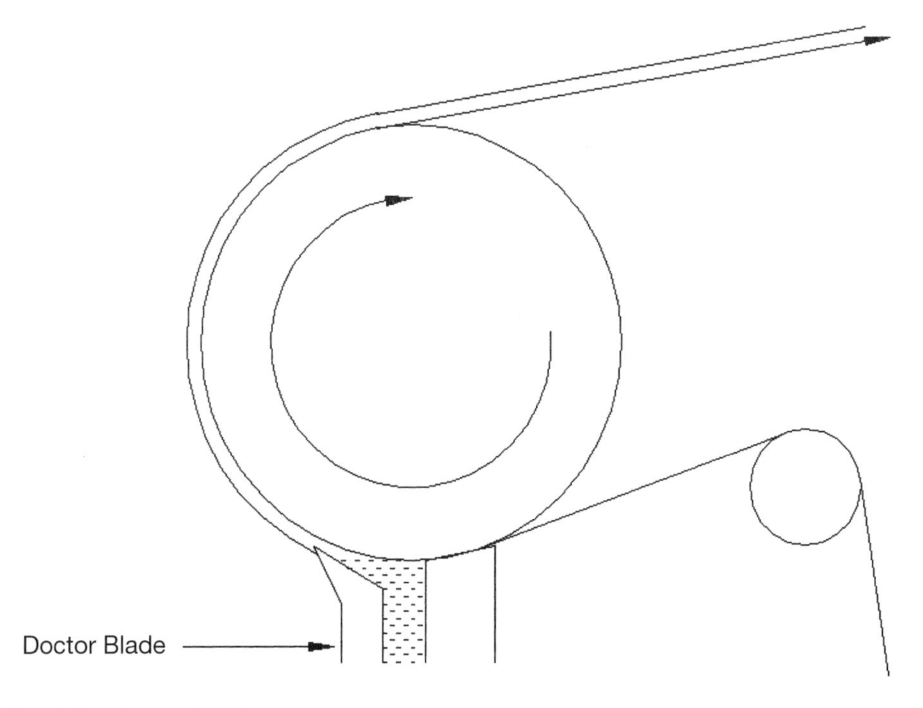

Fig. 4.8 Schematic of lip coating head.

system. As shown in Figure 4.10, the set position of the coating head can be switched between the on-the-roll configuration and the air cushion configuration, depending on the type of coating desired. The selection of which method to use is based on such considerations as web thickness or stiffness, desired coating thickness, and slip rheology. Heavier wet thicknesses on thin substrates are handled better on the roll. Thin applications (5 μm and less) are usually done off the roll. The slurry to be coated can be supplied to the die by either a gear metering pump or a pressure pump controlled by a precision gas regulator. The coating thickness is controlled by the pressure of the slip in the die and the speed of the web being coated. Manufacturers state that dry tape thicknesses in the range of 1 to 100 μm (0.000039 to 0.0039 in.) are possible. Slip viscosities in the range of 100 to 25,000 mPa·s (cP) can be used with the slot-die coating heads.

We have described a variety of coating heads for the formation of tape-cast materials. In doing so we have referred to carriers or substrates for the tape being cast. In the next section we will discuss the types of carriers available and the advantages and disadvantages of each.

4.1.3 Substrates or Carriers for Tape Casting

There are a wide variety of substrates or casting surfaces being used in laboratories and production facilities around the world. One of the

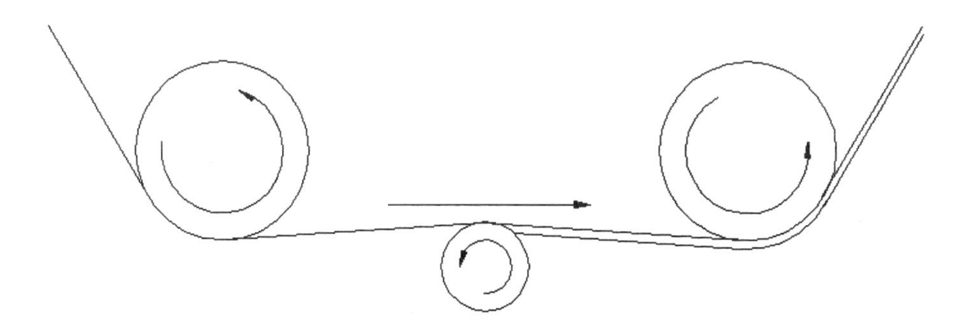

Fig. 4.9 Schematic of microgravure coating head.

most common casting substrates, especially in manufacturing operations, is a continuous stainless steel belt. This is by far the most cost-effective casting surface, since there is no polymer carrier to dispose of after the tape is dried and removed. Usually these steel belts are made of a nickel-chrome stainless composition and have at least a 2B, bright surface finish. The major drawback of stainless steel belts is that tape removal from the surface, especially with thin tapes in the 25- to 50-μm thickness range, is very difficult. Tape removal systems will be discussed in a section later in this chapter, but a scraping knife-edge type of removal blade is used with most stainless steel casting belts. If the knife blade is harder than the belt, then scratching is a distinct possibility, and the casting surface is ruined. Some manufacturers have circumvented this problem by using harder, more scratch-resistant belts made from carbon steels or valve

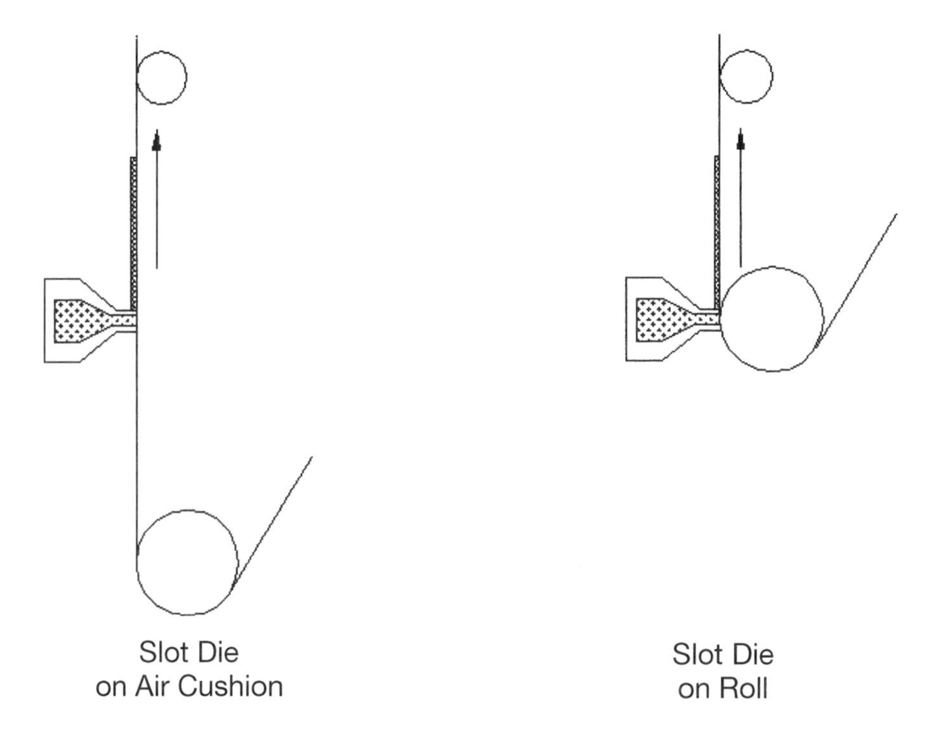

Fig. 4.10 Schematic of slot die coating head.

steels. At any rate, the steel belts do deteriorate with age and have to be replaced at fairly frequent intervals. This replacement process can be costly. Binder/plasticizer formulations must be developed specifically for use on the stainless steel belts, since some binders tend to stick more than others. Some operations use a release agent such as lecithin or other fugitive material that is dissolved in a solvent and sprayed onto the steel belt, dried, and buffed before the cast is made. A complete system for the application and buffing of the release agent can be built into the casting machine. Thin tapes still present a problem, since they lose quite a bit of their tensile strength when their cross-sectional area becomes so small.

Many laboratory casting machines are designed to use smooth glass plates as the casting substrate. Most of these machines are of the moving blade/stationary substrate variety. The machine designs and types will be described in a later section of this chapter. Once again, the binder/solvent/plasticizer selection for use on glass plates is critical, since many binders are very adherent to glass surfaces. One interesting scale-up of the glass plate casting surface reported in the literature is the use of a stationary blade with a continuous rotating glass circular plate machine.[5] The machine described in that paper has a glass plate diameter in the range of 1000 to 3000 mm (39 to 118 in.) and is therefore equivalent to a 3- to10-m (10- to 32-ft.) long tape caster. The glass plate is made by the float-glass process and is fairly flat and smooth.

Other very common carriers found in industrial applications are either polymeric or coated paper. The polymer carriers range from polypropylene to polyester and well beyond. Polypropylene is less expensive and is used in many of the capacitor manufacturing operations. The surface quality of the polypropylene is often not as good as that of the polyester carriers. The most forgiving casting surface from a release standpoint is a silicone-coated polyester. This casting surface works for most combinations of binder/solvent/plasticizer. By "works," we mean that the dried tape releases easily and does not interact with the carrier itself. Polyester carriers can be used at elevated drying temperatures up to about 100°C. This is another consideration when selecting a carrier, since not all polymer materials will maintain their mechanical properties at elevated temperatures and will tend to stretch during casting.

Other polymer carriers such as cellulose triacetate, Aclar™ (a fluorinated Mylar™), Teflon, and cellophane have been used for very specific applications. We have had excellent success with the silicone-coated polyester for a number of binder/solvent/plasticizer systems with a wide variety of materials. A cost-competitive and environment-friendly coated paper carrier has been evaluated and described by one of the authors in a recent publication.[6] These paper carriers can be recycled easily and yield tapes that are equivalent to the polymer carriers.

The best carrier for a particular application is determined by the interaction of the binder/solvent/plasticizer system with the carrier material. This determines how well the slip wets the surface and how well the dried tape releases from its surface.

During casting it is very important that the slip wets the carrier sufficiently. Recent studies[7] have shown that if the wetting angle is too high, the surface tension forces can cause internal phase separations. This in turn leads to green density gradients, which result in crack formation even in the early stages of drying. Most organic solvent–based slips wet the polymer carriers with no problems at all, but with water-based systems, wetting becomes a major factor. Most systems are evaluated on several carriers during development, and the least expensive one is selected for production. There are some applications for which several uses of a carrier can be made, and there are others, such as substrates for thin-film applications, where a single use is demanded.

A Case in Point

During the development phase of a new product that used tape casting for a very low-tech product—Christmas tree ornaments—one of the authors and his partner in the venture, Dan Shannon, decided to use an Aclar™ film for as many casts as possible. The carrier, after the product was stripped, had a residual film on its surface that had to be removed before it could be used for the next series of casts. We therefore decided to run the machine in reverse and rub the surface clean using ethyl alcohol and labwipes. This process required both of us, since the reverse speed on the machine was "full throttle" and we could barely keep up. As a matter of fact, we resembled a scene out of a well-known "I Love Lucy" episode in a candy factory. The entire process took several hours every week and in the long run was not very cost effective. The moral of the story is twofold: first, determine whether re-use of the carrier is going to save any money, and second, make sure the speed of the machine in reverse can be controlled!

4.1.4 Tape Casting Machines

Tape casting machines, like doctor blades, come in as many sizes and shapes as there are companies and designers producing them. They range from simple laboratory-scale machines to large production-scale machines with a wide variety of carrier transport and chamber designs. Examples of several of these machines are shown in Figures 4.11 to 4.16.

The basic principles of the tape casting process are included in most of these machines whether they are laboratory casters or production-scale models. The major components needed for tape casting include: a solid, level casting surface, a drying chamber with a built-in means for controlling the airflow over the drying tape, an adjustable speed carrier drive control with a constant speed regulation mechanism, an air heater to control the temperature of the filtered feed air to the drying chamber, and underbed heaters to set up the desired temperature profile in the machine. Most machines also include a master control panel that contains all of the instrumentation in one convenient location. Not all of the machines shown in Figures 4.11 through 4.16 contain all of these elements. For example, the laboratory caster shown in Figure 4.12 does not have underbed heaters, since it is designed to be a batch caster for very short experimental runs of tape-cast product.

Casting machines are very simple in design, since they are really only elongated (in some cases) forced-air drying ovens. The addition of multichambered drying with separate temperature controls (and at times separate airflow controls) can produce a wide variety of temperature-time profiles in any given machine. When designing a casting machine, one must keep in mind that its primary purpose is to

Fig. 4.11 Production slot die tape caster, ceramic capacitor coater, Model OMD-S/3.5/D/8.5H, Photographs courtesy of Yasui Seiki Company, (USA), Bloomington, Indiana.

remove the solvent(s) as quickly as possible without forming a skin on the surface at too early a stage of drying. The creation of a skin, or "case hardening" effect, can occur if the solvent gradient and/or the temperature gradient is not controlled or adjusted properly. The drying process is discussed in more detail in a later section in this chapter. The ideal situation in a tape casting machine is to have a high solvent vapor concentration at the entrance end of the machine near the casting head and a continuously decreasing concentration as you pass down the casting chamber toward the exit end. This is usually accomplished by placing the exhaust fan or duct just downstream from the casting head assembly. Some machines use a baffle system to accomplish the same effect.

The ideal temperature gradient in a machine is to have room temperature at the casting head and a continuously increasing

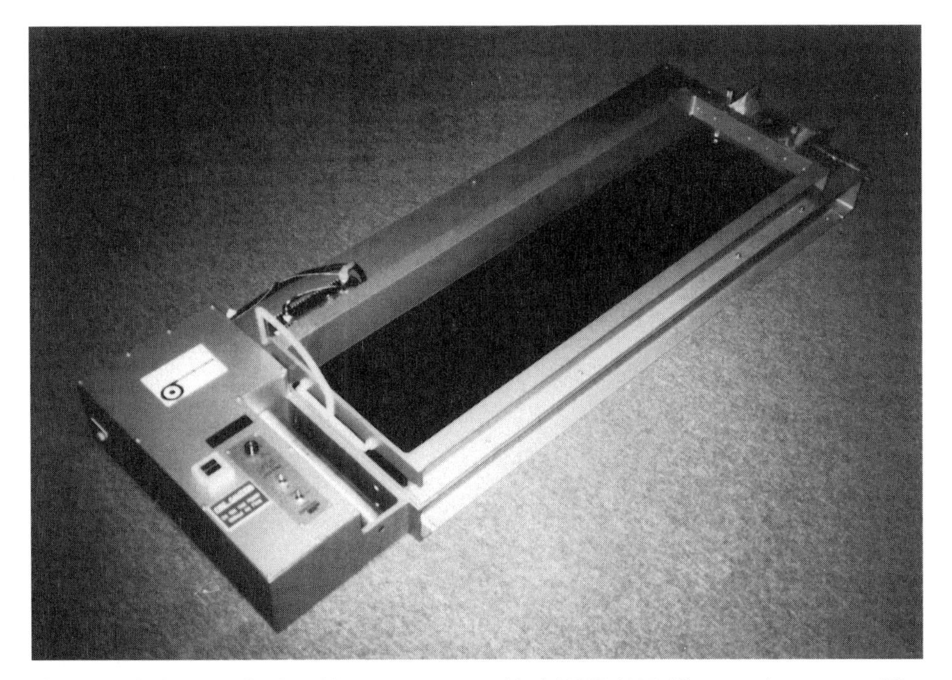

Fig. 4.12 Laboratory Scale, table top tape caster, Model TTC-1000, Photograph courtesy of the Tape Casting Warehouse, Morrisville, PA, 19067.

temperature as you approach the exit end of the machine. This is usually accomplished by situating the air heater assembly near the exit end of the machine. Most machines have a "dead" space designed into them beyond the heater so that the dried tape can cool down before removal or rolling onto the carrier. The peak temperature in any tape casting machine should be reached at a point where the majority of the solvent has been removed from the then-almost-dry tape. On some machines there is also a "dead" zone just downstream from the casting head assembly. This zone is incorporated to prevent the rapid drying and the resultant skinning effect mentioned earlier. We will now describe and define some of the major components of the tape casting machines in use today.

4.1.4.1 Casting Surface

We have mentioned the importance of a flat, level, and smooth casting surface. The surface under the carrier is the ultimate gauge of the thickness uniformity that can be achieved during casting, regardless

Fig. 4.13 Dielectric ceramic tape casting machine. Photograph courtesy of A.J. Carsten Co., Ltd. Powell River, BC, Canada.

of the type of casting head used. If the casting surface under the carrier is not flat and smooth, the final tape-cast product will not be flat and smooth. The ultimate flatness is obtained from the slot die coater, because coating occurs "off-roll." In this case the flatness of the carrier and thus of the coated film is controlled by the tension on the carrier, since there is no hard surface under the moving film. In most other coating heads, including the doctor blade process, there is a solid base of some sort under the blade, roll, or slot die. In some cases the casting is done using a knife or blade on a plate of some sort. This plate can be metal, glass, or other material. In ideal situations, the plate under the knife or blade is a polished granite block. The use of a granite block provides a very flat ($< 1\ \mu m$ or 0.000039 in.) precision-ground and polished surface as well as a very solid and level casting bed.

Fig. 4.14 Pro-Cast® tape casting machine. Photograph courtesy of Unique/Pereny Division of HED Industries, Ringoes, NJ, 08551.

If a steel belt is used as the casting surface, it becomes the casting bed as well as the carrier. This can create problems as the steel belt ages and develops surface scratches and ripples in its surface. Most manufacturers using steel belts have to replace them on a regular preventive maintenance basis to prevent problems with surface defects and thickness variation. Casting heads that include a roller under the carrier under the slurry metering system are very common on Japanese machines. Precision-ground steel rollers that rotate in the direction of the moving carrier system are used on these machines. The use of a curved surface under the carrier also insures that the carrier is not curled across the casting direction when it passes under the coating head. There is at least one tape casting machine that uses a precision-ground stationary roller under the polymer carrier—that is, it does not turn as the carrier passes over it. The manufacturer claims that the fixed roller system eliminates roller diameter tolerance variation.[8]

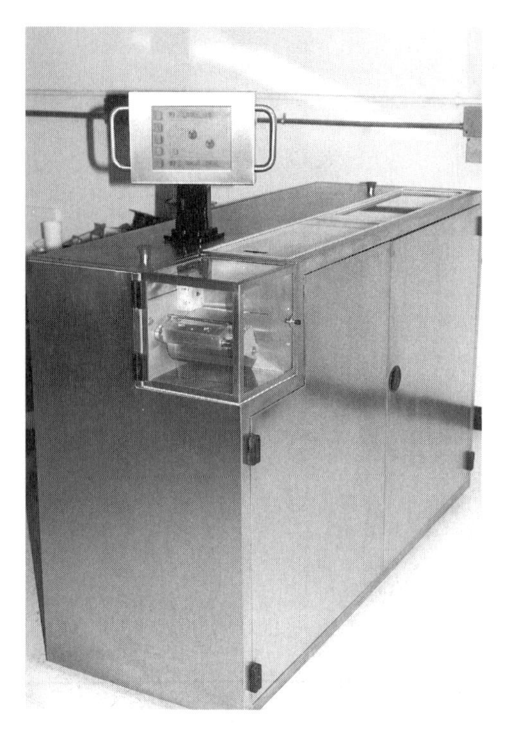

Fig. 4.15 Dreicast coating system, Model 1000-001. Photograph courtesy of DreiTek, San Marcos, California, 92069.

4.1.4.2 Support During Drying

Once the film is metered onto the carrier surface it is conveyed into the drying chamber, where the actual solvent removal takes place. For a period of time, depending upon the thickness of the cast and the drying rate, the fluid film on the surface of the carrier is free to flow. Therefore the surface it rides on must be flat, smooth, and level. During the first stage of drying, the fluid suspension of particles in the slip forms a semi-solid, gel-like structure. When this gel forms, the film becomes more stationary and the requirements for the underbed support become less rigid. For very thin films of slip, such as those for ceramic capacitors (in the 25-μm or 0.001-in. and less range), this gel formation process is almost instantaneous, and in some cases no underbed support is required at all. Some machines are designed with the casting surface and the casting bed contiguous in the drying chamber—that is, they are made from one continuous piece of material. These are usually laboratory-scale machines such as the one shown in Figure 4.12. Several of these laboratory machines use flat glass plates made by the float-glass process, and others use a polished granite block.

In some production machines that use a knife-on-plate type of casting head and a granite block casting surface, the granite block is extended into the drying chamber through the "dead zone" to the exhaust fan exit. This extension of the uninterrupted flat, smooth, and level zone allows the cast film to form a gel-like structure before it reaches the heated drying chamber. Beyond this point, heavy-

Fig. 4.16 Continuous style tape caster. Photograph courtesy of EPH Engineering Associates, Inc., Orem, Utah, 84058.

gauge aluminum plates, which are excellent conductors of heat, are used to form the drying chamber bed through the remainder of the machine. On machines that use a continuous steel belt as the carrier, the steel becomes the moving bed that rides on precision-ground rollers. There is also a bottom plate under the rollers, which creates a closed drying chamber around the entire conveyor system.

Some production machines use a synchronized moving polymer loop carrier as the support under the polymer casting substrate. This polymer carrier transports the tape-cast film on its polymer substrate through the dryer with a minimum amount of friction. The transport carrier rides on a series of precision-ground rollers through the drying chamber (or chambers) in the same manner as the steel conveyor.

The latest designs have eliminated the carrier polymer loop, and the casting polymer carrier is transported through the drying chamber on precision-ground rollers. Some other drying zones use a fluidized-bed type of conveyor system, with the heated air from the top and bottom providing a cushion so that the carrier actually floats on the air column through the drying chamber. Precise tension control has to be maintained in this system to prevent "flapping" of the polymer in the drying chamber. Tension control is maintained using "dancer rolls" and clutching systems to provide the low tension necessary for these thin carriers (sometimes less than 50 μm or 0.002 in.).

A unique support system was described in the literature[2] that uses a curved bed of smooth tempered glass. The glass bed was bent into the arc of a circle with a radius of 132 m (430 ft.). The entire glass casting bed was a single piece of glass that was 4.3 m (14 ft.) long and 305 μm (12 in.) wide. The curvature was provided to assure close contact between the thin carrier tape and the bed and to insure that the tape did not have a tendency to edge-curl during drying. The maximum height of the circle was located under the doctor blade so that the surface at this point would be horizontal in the longitudinal and lateral directions, in order to prevent any tendency of the slip to flow under gravitational forces. The authors reported using this casting machine to produce lead zirconate–lead titanate tapes in the thickness range of 25 to 300 μm (0.001 to 0.012 in.) with a thickness tolerance of ± 7.62 mm (± 0.0003 in.).

Tape casting machines have been built that include a mechanical restraint system for holding down the edges of the polymer carrier during the drying process in order to prevent edge curling. These mechanical restraints range from simple slots in which the polymer carrier edges ride to "tank treads" that ride on the two outside edges of the carrier. We believe that the proper solution to the edge curling problem is in the design of the chemistry of the system. This will be discussed in more detail later in Section 4.3.

4.1.4.3 Drying Chambers

Drying chambers come in a variety of sizes and shapes. Some of the continuous casters have a small cross-sectional area on the order of 100 × 300 mm (4 × 12 in.), while others use an open box where the carrier takes a circuitous path through the drying chamber while passing over heated plates and drums. Often the purpose for these box-type chambers is to reduce the final length of the casting machine. A 10-m (32.5-ft.) machine can be "squeezed" into a 2-m (6.5 ft.) long box. We feel that the small–cross-section continuous casters provide the best drying conditions by producing the desired solvent and temperature gradients necessary, especially for the thicker tape-cast products (> 0.254 mm or 0.010 in.). Some of the small–cross-section machines are really not that small. We have been involved in the design of machines with cross sections as large as 100 × 1270 mm (4 × 50 in.). These machines had multiple drying zones and were over 30 m long. Thinner tapes, such as those now being manufactured for multilayered ceramic capacitors (MLCC), are much more forgiving since they are essentially all "skin," and drying problems are all but eliminated. These tapes can be manufactured in the open-box type of casters. The major requirement in the production of these tapes is to get enough heat and airflow to the tape fast enough to meet production rates as high as 30 m/min (97.5 ft./min). The variety of casting machine designs, especially for the drying chambers, can be seen in Figures 4.11 to 4.16.

Most, but not all, of the production machines have a series of chambers where the temperature and airflow characteristics can be tailored to the proper drying conditions for the tape being cast. A typical 8-m-long drying chamber may be divided into three zones with lower airflow and temperatures in the first two zones and much higher airflow and temperatures near the exit end. Machines have been built with a wide variety of airflow controls ranging from countercurrent airflow (to the direction of the cast tape movement) to concurrent airflow. The flow of the drying air through the chamber over the cast tape is critical to the tape casting process.

4.1.4.4 Air Flow

We have already mentioned that countercurrent airflow is preferred for maintaining a solvent and temperature gradient throughout the drying chamber. The velocity of airflow is also critical to this process. In a previous publication,[9] a countercurrent laminar airflow during the drying process was recommended for solvent-based systems containing trichloroethylene and ethyl alcohol. It was found experimentally that an airflow of greater than 3.1 m^3/hr (110 $ft.^3$/hr) would cause severe skinning on the tape-cast product. The volume of airflow that can be tolerated by any tape casting system during drying is predicated on several factors. These include: the type and concentration of solvent(s), the solids loading in the slip, the type and concentration of binder, the temperature of the drying air, and the thickness of the wet layer (and correspondingly the dry tape thickness). In addition, the early stages of drying cannot tolerate as high a flow rate as the final stages of drying. This is the reason many of the production machines are built with a series of drying chambers, each with its own air velocity and temperature control. Many of the machines use air bar dryers, which form a vortex countercurrent airflow upon which the carrier film rides during the drying process. The vortex airflow prevents flapping of the carrier and provides very high efficiency of heat transfer during transport through the machine. This principle is shown in Figure 4.17.

Another factor in airflow requirements during the casting of any flammable solvent(s) in the presence of heat is safety. There are published regulations as to the volume of airflow that must be maintained to remain below the lower explosion limit (LEL). Each solvent system should be studied before using a specific casting machine to determine whether the machine will be operating in a safe zone with respect to the LEL. Information about these calculations can be found in the published literature.[10] Further discussion of this topic will be included in Section 4.3 and in Appendix #2.

The airflow through the drying chamber can be controlled by several techniques. The size of the exhaust fan (if the air is being drawn through the drying chamber) or the size of the blower fan (if the air is being force-fed into the drying chamber) with respect to the volume of the drying chamber is one such factor. The rotation speed of the fan is another. In some cases a variable-speed motor is used to control the airflow through the machine. A much simpler and less costly system is to use a series of dampers that partially divert the

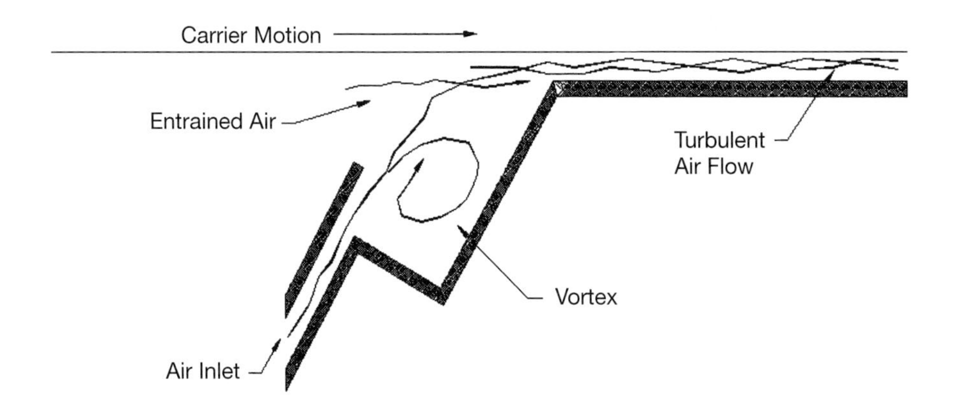

Fig. 4.17 Airbar vortex floating dryer schematic.

airflow from the drying chamber. One or more of these dampers can be built into the machine to provide the airflow necessary to dry the tape-cast product using the minimum airflow necessary to maintain safe operating conditions. Many of the production machines use a series of hot-air circulation fans along the drying chamber. These fans are usually mounted above the chamber. The heated air is blown downward through a series of baffles to finally impinge upon the drying tape, while the flow of solvent-laden air is directed toward the edges (or sides) of the tape and then back up into the recirculating fan. This type of system does not provide the countercurrent airflow that we feel is necessary for drying thicker tapes, but it works efficiently for the thinner capacitor tapes being produced today (those in the 25 μm (0.001 in.) and less range).

Another consideration to be addressed is whether the drying chamber is under positive or negative pressure. If an exhaust blower is being used to pull the air through the machine, then the drying chamber is considered to be in a negative pressure state. Any leaks would involve airflow into the machine, not out of it. On the other hand, if the air is being blown into the drying chamber, as in the case where house air is being used, then the chamber is under positive pressure, and any leaks would be out of the machine into the surrounding atmosphere. There are benefits to each type of system. The obvious benefit for a negative-pressure system is that the solvent vapors remain in the machine and are carried out in the exhaust stream. There is a possibility of drawing particulate contaminants into the drying chamber from outside the machine, and if this is a problem, then the machine has to be installed in a clean room atmosphere. The positive-pressure drying chamber presents exactly the opposite picture. There is the possibility that solvent vapors will be expelled from the drying chamber into the room, and therefore the machine must be under an exhaust hood of some sort. On the other hand, when a machine is under positive pressure there is little chance of particles from the room contaminating the drying tape. Today's machines use either the positive or negative pressure system, and they work well when the proper precautions are taken.

Some of the larger production machines built in the United States—that is, those that exceed 30 m (100 ft.) in length and 1.27 m (50 in.) in width—have been built with a series of chambers, each of which uses air inlet sparging bars and exhaust bars at the opposite end. These systems provide for the countercurrent airflow over the drying tape and also are quite readily adjustable using damper valving mechanisms at either the inlet or the exhaust or both.

All fans used on or near the tape casting machine should be explosion-proof if volatile, flammable solvents are being used in the process. On many of the laboratory-scale machines this problem is eliminated by the use of "house air" provided by a remote compressor. On these systems, the airflow through the machine is controlled by the use of flow meters and the cross-sectional geometry of the drying chamber itself.

Currently, many of the tape casting machines being built incorporate an air filtration system to prevent dust and dirt from contaminating the wet tape surface. This is particularly important for the thin capacitor tapes that are almost as thin as or thinner than the dust particles. In these machines, all of the air supplied to the tape casting chamber passes through a HEPA (High Efficiency Particulate Arresting) filter that removes approximately 99.97% of all of the particles present in the room air. Since many of these machines operate in a clean-room atmosphere of Class 1000 or better, there are few particles present for the filter to remove. The HEPA filters come in a variety of particle removal efficiencies. Filters should be selected according to smallest-sized particle that can be tolerated as a potential defect and the airflow restriction that may result. Consult the manufacturer of the tape casting equipment during the design phase so that the proper filter can be installed. In any event, these filters can be interchanged, and a different efficiency filter can usually be substituted at a later date.

4.1.4.5 Heat Sources

Almost all production tape casting machines and many of the laboratory machines incorporate some sort of heater to raise temperatures during the drying process. As mentioned previously, the ideal situation in a tape casting machine is to have the entrance end of the machine at or near room temperature and the last zone before the cool-down zone at or near the boiling point of the solvent system being used in the tape formulation. For most systems in use today, including water, the maximum temperature in a tape casting machine will be less than 100°C. There are several different techniques for providing the heat necessary for the accelerated drying of the tapes and for providing the temperature gradient just described.

Most tape casting machines provide for heated airflow in the casting chamber. The simplest system is to draw or force the air through a heater of some sort. These are usually resistance-type heaters with a control thermocouple in the airstream just downstream from the heater. Almost all of the air heaters are also equipped with an overtemperature control thermocouple to provide for failure of the main control thermocouple. In addition, the air heater is usually provided with an automatic shutdown system in the event of fan failure. This eliminates the danger of overheating elements and the potential for fire or explosion. On machines in which there is one air heater, it is mounted at the exit end of the machine so that the heated air can be drawn through the machine countercurrent to the movement of the drying tape in the machine. A series of temperature-monitoring devices, either thermocouples or thermometers, are usually mounted along the drying chamber to observe and record the temperature profile in the flowing hot airstream. On some machines, where there are a series of independent (although we dispute whether they are truly independent) drying chambers, the recirculating fans have individual heaters. Each of these recirculating fan systems is provided with a thermocouple temperature control. These fans are of the blower type and are forcing the hot air into the chamber, thus creating a positive pressure in the chamber. On systems where the fan is at the exhaust end and where the air is being pulled through the machine, the drying chamber is under a slightly negative pressure.

There are also large-scale production machines in which the heated air is provided from a single source into a series of chambers with different flow rates controlled by damper-and-valve systems. Each manufacturer has its own techniques for providing the temperature and airflow control in the machine chamber. Many of the smaller laboratory and production machines control the heated airflow inside the entire drying chamber by providing for a balance between an inlet air blower and the exhaust air fan. On at least one laboratory machine that uses compressed "house air," an in-line heater is provided as an option to produce an elevated temperature inside the casting chamber.

In addition to the heated air option, most tape casting machines, especially those used for full-scale production, use underbed heating to accelerate the drying process. The underbed heaters heat the bottom of the carrier, which increases solvent diffusion up through the gel structure to the top surface. The heated airflow over the surface of the tape then evaporates and sweeps away the solvent vapors to complete the drying process. Usually the underbed heaters are arranged in the casting chamber in a series of zones, each of which can be individually monitored and controlled electronically. A typical tape casting machine that is 15 m long could have as many as 10 separate controlled heating zones. By balancing the heated airflow over the drying tape with the controlled underbed heating zones, manufacturers can create an exact temperature profile along the length of the machine. Separate monitoring thermocouples should be in the drying chamber close to the drying tape itself so that the exact drying profile can be recorded. Underbed heaters are usually not used where the heated airflow passes both above and below the tape on the polymer carrier. Systems where the carrier passes through the drying chamber on rollers or on air flotation usually have only heated-air dryers. The use of underbed heaters is also usually not recommended for steel-belt carriers due to heat diffusion limitations.

There are also instances where the use of overhead heating is warranted. In those situations, radiant heaters are mounted in the roof of the drying chamber. These radiant heaters are also electronically controlled and are situated to provide additional drying at the exit end of the drying chamber.

4.1.4.6 Conveyer Controls

One of the most important systems on a tape casting machine is the carrier or conveyer speed control. The accurate control of the conveyer speed while it is passing under the coating head is one of the variables that determines the final coating thickness on its surface. If a continuous web or belt is used under the polymer casting substrate, it is relatively easy to maintain a constant speed for the casting operation. When the carrier is being rolled up onto a take-up core, even if the product has been stripped from its surface, constant speed control becomes more difficult, since the diameter of the take-up increases with time. This, in turn, increases the speed of the carrier passing under the coating head. On most tape casting equipment that uses a take-up roll, the speed is controlled by a tachometer-type speed pickup, which rides on an "idler" roll, that is, a roll that rotates as the carrier moves on its surface but is not a driven roller. With a digital feedback speed sensor of this type, it is possible to maintain carrier speed control to within 1% of the set-point speed. With machine casting speeds ranging from 12.7 cm/min to more than 50 m/min (5 in./min to > 162 ft./min), the precise control of the speed within specified limits becomes very important. When casting at the slower speeds, the casting speed uniformity becomes much more important.

Tension control works in concert with the speed control on a tape casting machine. A wide variety of tension control mechanisms are used, ranging from a very simple spring-loaded frictional strap on the payout roller to clutch-and brake-systems that are electronically activated. The simple friction-based systems are manually adjusted, whereas the electronic systems use a tension pickup and feedback loop to modulate the tension on the web. On some of the larger-production scale machines, especially those that operate at speeds over 10 m/min, intricate dancer rolls and multiple tension-control systems are used at both the payout and take-up ends of the machine. Each manufacturer has its own system for conveying the carrier through the drying chamber. We believe in the KISS principle (Keep It Simple Stupid) in the design of conveyer systems for tape casting. The more complex the system, the more that can (and usually does) go wrong, and the harder it is to set up the machine for each casting run. Tension control is important, but there are machines on the market that tend toward overkill and drive up the cost of the machine.

Another conveyer control system that is important, especially when casting continuously on a polymer carrier, is precise edge alignment of the polymer as it is wound onto the take-up roller at the exit end. These control systems are electromechanical—that is, the edge of the carrier is detected electronically and a set of rollers moves back and forth as the carrier weaves over and around them. The mechanical movement of the rollers can keep the carrier centered on the take-up roller with less than ± 3 mm (± 0.118 in.) lateral displacement. This same kind of power tracking and centering system can be used at the polymer unwind station to assure precise control and location of the carrier as it passes under the casting head. Most of the large-scale production machines have one or both of these tracking systems.

Another feature of the conveyer systems is static control. Almost all polymeric carriers generate static electricity during winding and unwinding. We have been "zapped" many times by static discharge. It is important to remove this static charge for at least two reasons: static attracts dust particles to the surface of the carrier, and static discharges in the presence of flammable solvents can cause an explosion. Static eliminators come in a wide variety of forms, ranging from a simple system of stainless steel grounding brushes in contact with the back side of the carrier (the side opposite the casting surface) to ionizing discharge systems that reduce or remove the static electricity on the carrier web. It is highly recommended that some sort of electrostatic discharge system be installed on a tape casting machine with a moving polymer carrier. Most manufacturers install these systems as a standard part of the tape casting machine.

4.1.4.7 Slitters, Cutoffs, and Take-Up Systems

Once the tape-cast product is dry and ready to be removed from the machine, there are several procedures that are sometimes used to facilitate further processing downstream from the forming process. One that is almost always present is an edge-slitting process, which removes the thin section (sometimes referred to as a "feathered" edge) on each side of the tape-cast web. The edge slitter consists of a hardened steel, precision-ground support roll and a disc blade that cuts through the tape. The thin section, which is about 6 to 12 mm (0.25–0.5 in.) wide is discarded or recycled, and the uniform-thickness tape in the middle is the final product.

This same type of slitter is also used to cut a wider piece of tape into narrower sections at the exit end of the machine. For example, a 200-mm-wide tape may be slit into four 50-mm strips. Each of the 50-mm strips would then be rolled as a separate entity onto the take-up roller(s). Another type of slitter used to cut a wide tape into sections is based upon the scissors principle—that is, it has two cutter wheels or block knives that overlap each other and shear the tape like a pair of scissors. The block knives on each of these systems are driven and are synchronized with the movement of the tape-cast product. When a wide tape is slit into two or more narrower widths, there are usually two take-up axles. As the slit tape moves from the slitter, alternating strips are diverted to different take-up axles, thus eliminating the possibility of overlap or interference between adjacent rolls.

If the tape is to be slit into narrower strips, the carrier is almost always removed from the tape prior to the slitting operation. This is usually accomplished by having the carrier make a 180° bend around either a knife edge or a roller. The tape-cast product continues in a straight path once it is stripped from the carrier, and the bare carrier is rolled onto a take-up roller. The tape-cast product then passes into the slitter section.

There are situations where the tape-cast product is removed from the carrier and is cut into sheets rather than being rolled. An example of such a situation is in the production of large sheets for the manufacture of molten carbonate fuel cells. Some of these sheets are as large as 1.2 × 2.4 m. These tapes are usually edge-slit to remove the "feathered" edges and are cut to length using a "flying blade" product cutoff. The blade is a hardened steel disc that is time adjustable, automatic, and air activated to roll across the tape at precise intervals to cut the tape into sheets of a predetermined length. The blade movement can be synchronized with the tape movement to provide as close to a 90° cut as possible. On some production machines a simple knife cutoff is used where the operator manually cuts the tape with a razor blade. The sheets that have been cut are then removed from the take-up table, and the next piece moves into position for cutoff.

If the product is to be rolled at the end of the casting machine, there are two choices: the product is either stripped from the carrier or it remains on the carrier for ease of handling downstream from the casting operation. If the tape-cast product is to remain on the carrier, it usually is wound onto either a cardboard or a polymer core, which is held in place using quick-release chucks. These chucks are used to facilitate rapid changeover during a casting run. A roll of tape on a core can be removed and a new core can be put in place on the fly without shutting down the process. The take-up roller is driven and is synchronized with the tachometer control system described previously to provide for a smooth and constant speed of casting.

If the product is being stripped from the carrier, the take-up system is a bit more complex. The speed control to provide constant speed during the cast is actually on the take-up roll for the carrier, and this speed is maintained by the tachometer feedback loop described previously. The product, on the other hand, must be handled with a tensionless take-up system, since there is no carrier to provide the mechanical strength needed in a system under tension. This is accomplished by using a proximity sensor with a loop of tape where a servo-motor accelerates and decelerates the take-up roller, depending on where the loop is with respect to the sensor. This system provides a tensionless take-up system for the tape-cast product. Many different types of proximity sensors have been used in these systems. The main requirement is that the sensor detects the tape that is being speed controlled. At times, especially when the tape-cast product "blocks"—that is, sticks to itself—a paper or other material must be interwoven between the layers on the roll. Machines have been built with all of these features added as options.

4.1.4.8 Other Machine Options

This section will describe some of the other options that are sometimes included on tape casting machines. The options in some cases are more expensive than the machine itself. All of the options described in the following paragraphs are usually found on large-scale production machines, where tight control of the process is required or where there is a concern about the volume of solvent(s) being generated in the exhaust stream.

Many tape casting machines are equipped with a wet-thickness monitor of some sort. These instruments range from a simple point micrometer, where the thickness is checked manually by the operator just downstream from the doctor blade, to very sophisticated noncontacting scanning gamma backscatter thickness monitors. The gamma backscatter technique is based on the principle of Compton photon backscatter. A radioactive isotope in the sensing probe emits low-energy gamma rays, which are collimated and beamed at the material to be measured. The gamma rays are scattered back toward the detector in direct proportion to the mass of the material in front of the probe. A scintillation crystal/photomultiplier detector converts the backscattered photons to an electrical signal, which in turn can be related to weight-per-unit-area or thickness (if the material density is constant). This output is then recorded and/or displayed for the operator. The instruments come in two modes: (1) the scanning mode, where the probe moves across the cast tape and reads the thickness variation from side to side, and (2) the in-line mode, where the probe is stationary and records the thickness variation along the cast tape at one lateral position (usually the center). There are a wide variety of thickness monitors on the market today, some of which utilize X rays or other radiation sources. The gamma backscatter system is one that we have used with excellent success.

Air pollution control is another option that appears on several tape casting systems, especially those that emit volatile organic compounds into the exhaust waste stream. There are several potential solutions to this problem. The first of these is to eliminate the use of VOCs altogether. Many companies are trying to do this by using water-based systems. Water-based technology is the subject of another section in this book. If VOCs are used in the process, there are several abatement systems that will provide a relatively clean exhaust stream. The simplest and least intrusive to the drying process is the use of an afterburner or incinerator on the exhaust stack. The burner, which is similar to a jet engine, is mounted at the top of the exhaust stack and burns the solvents being emitted. The burner can be gas or electrically operated, and it works in concert with the airflow through the tape casting machine. There is usually a high initial equipment cost involved, and the process itself is energy intensive.

A second option is a solvent-trapping system that condenses the solvent vapors into liquid form. This is an efficient type of solvent-removal system. It works by extracting heat from a vapor stream to cause the solvent(s) in the stream to condense to a liquid.

Solvents are sometimes recovered in a reusable and recyclable form. Condensation is a very safe system and presents little or no fire hazard. It is an excellent system for continuous process control and has a fairly low operating and maintenance cost. Systems are available for airflows of less than 28 m^3/min (708 cu.ft./min), which is well within the range needed for tape casting and drying.

Another type of system used in manufacturing is based on carbon adsorption. The exhaust stream passes through a carbon bed and the VOCs are adsorbed by the carbon particles. This is one of the most cost-effective systems, but it only works well with very low concentrations of VOCs in the stream, and the efficiency tends to decline with use. As the carbon bed becomes filled with solvent(s), it may also experience bed plugging, which decreases removal efficiency and can present a fire hazard. In addition, back pressure can cause problems with the airflow system in the tape casting drying chamber. The carbon bed must be periodically reactivated either by firing or by steam stripping, both of which can be costly maintenance operations.

Programmable logic controllers (PLC) interfaced with a personal computer (PC) have been introduced on many of the complex tape casting machines to provide for precise control and monitoring of the tape casting process. From the single PC station the machine operator can monitor and make changes to any facet of the process, including heater set points, web speed, web tension, and dryer airflow. The monitor provides information about the machine status as well as the alarms and any problem areas it may detect. The use of a PLC/PC system eliminates the need for multiple temperature controllers (one for each zone of heat). In addition, the PLC/PC can collect data from any run and process it for future analysis.

4.1.5 New Tape Casting Technology

In this section we will discuss a relatively new technique being developed in laboratories around the world.[11,12,13,14] It is a solvent-free tape casting process that relies upon the use of ultraviolet (UV) curable binder systems. The process for forming the thin films on a carrier surface is identical with that described previously. Only the drying, or in this case curing, step is modified to produce the flexible tapes that are ready for further processing at the end of the casting machine.

All of these UV-curable systems have very different chemistries than those used in solvent-based systems. Most contain a dispersant, a plasticizer (not necessary in all cases), a photopolymerizable binder, and an initiator to activate the UV curing. A typical binder/initiator system found in the literature[11] is a polyester acrylate binder and an initiator such as 2-hydroxy-2methyl-1-phenyl-propan-1-one. The binder is a liquid low-viscosity monomer diluted into hydroxy-ethyl methacrylate. Most of the UV-curable systems contain an acrylate monomer of some sort, since they have relatively low viscosities (100 mPa·s or cP) and excellent reactivity with the UV radiation. The low viscosity allows the preparation of ceramic slurries with a relatively high solids loading, and the resulting high-density green sheets have excellent mechanical properties.

The slurry preparation is very similar to that used for conventional tape casting. There is a dispersion milling and mixing procedure, followed by de-airing and finally by the tape casting process. Tape casting is done by conventional techniques, using a doctor blade in most cases. The slurry is cast onto a moving Mylar film, which then passes into the UV curing chamber. The UV cure chamber replaces the solvent evaporation step of solvent-based systems. During UV curing, the slurry on the carrier passes under UV lamps with output spectra in the range of 200 to 450 nm, with a peak intensity at 365 nm. In the example cited,[11] the average UV energy concentrated on the green sheet was about 450 mW/cm^2. A typical flowchart for the preparation of alumina sheets taken from reference 11 is shown in Figure 4.18.

The photopolymer mixed with the initiator polymerizes when exposed to the UV radiation, and the ceramic particles are entrapped in the polymeric matrix. Herein lies one of the advantages of UV-cured tape casting: a much shorter tape casting machine can be used since there is no long-term evaporative drying involved. Also, the elimination of solvent(s) means that solvent recovery or burnoff systems are not needed.

There are potential disadvantages to UV curing systems. The depth of cure due to limited UV penetration can create a problem with thick tapes. Most of the work reported to date has been up to about 350 μm (0.014 in.) in thickness. There are also some materials that limit the depth of cure. Some of the heavy metal compounds based on barium or molybdenum have limitations with respect to UV cure depth and can only be used for very thin tapes. This is not a problem for the thin tapes used in multilayered capacitors (as thin as 5 μm).

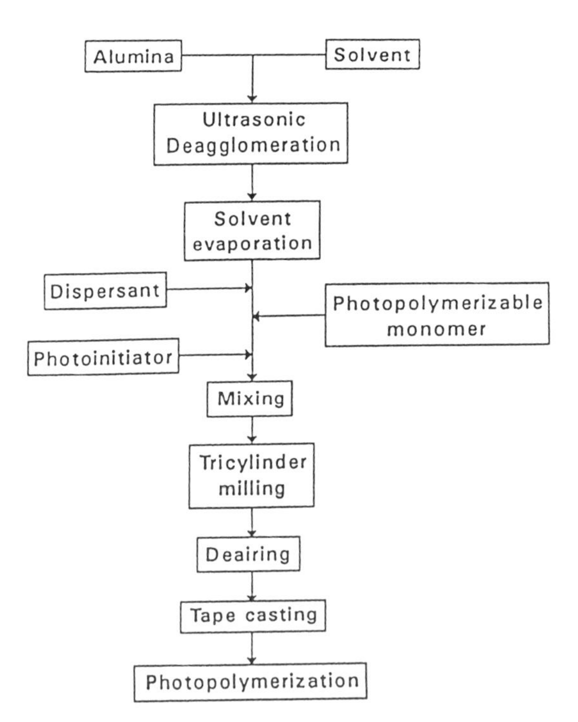

Fig. 4.18 Flow chart for UV curable tape casting (From reference 11).

The potential advantages far outweigh the disadvantages, and the development work is continuing in laboratories at many universities and some companies. To the best of our knowledge, there is no large-scale production system in use today for UV-curable tape casting.

Section 4.2 will discuss the theoretical aspects of the casting process, including the physics and the procedures for fluid metering.

4.2 PHYSICS AND PROCEDURES FOR FLUID METERING

Metering the fluid into a uniform layer is the crux of the casting process. More development work has been done on this step than on all of the other steps combined. The mechanical laydown of the film is accomplished in a myriad of different methods as was described in the preceding section on coating equipment. This section of the book will look at some of the forces that come into play during the coating process. The explanations and discussions throughout this section are based mainly on the doctor blading process. Unless specifically explained otherwise, it is safe to assume that all theories and conclusions are based on the doctor blading process.

The doctor blading process is essentially a wiping process in which the prepared slip is deposited on a carrier surface in excess of the amount needed and the excess is continuously wiped back off the carrier. The removal (wiping) of the excess material can be accomplished by either moving the wiping blade across a stationary carrier or by moving the carrier underneath the stationary wiping blade. Common sense will easily show that the stationary blade and moving carrier surface is much more easily scaled up into large production areas and volumes. The film left behind after the wiping, or doctoring, procedure is then dried to become the tape. The drying process and the resulting tape are discussed in other sections in this chapter.

The mechanical setup of the doctor blade tape casting process is shown in Figure 4.7. In this process, the slip is deposited in excess on the carrier surface, typically in a confined reservoir. The carrier surface is then moved under the doctor blade, which wipes the excess material back into the reservoir, leaving a thin layer of fluid on the carrier. The mechanics of the doctor blading process are, in theory, quite simple. In practice, however, controlling all of the variables to generate a uniform film can be quite difficult. The list of variables that affect tape uniformity is fairly long; it includes slip homogeneity, viscosity, fluid nature (that is, flow behavior or rheological property), slip surface tension, slip-carrier wetting behavior, wiping speed (casting speed), blade height (gap), amount of excess material (reservoir height), and blade size/shape. Some of these variables, such as blade size and shape, are easily held constant. Others, such as casting speed, are mechanically controlled within a tight range, while such factors as slip homogeneity are much harder to measure and therefore to actively and accurately control. Each of these variables will be addressed here, although the overlap between variables can be quite extensive. There are four primary defining variables for wet tape thickness, each modified by other secondary factors. The four primary variables are blade gap, casting speed, viscosity, and reservoir height.

4.2.1 Slip Homogeneity

The homogeneity of the cast film, or cast tape, is directly related to the homogeneity of the fluid being cast. On a bulk scale, a 20 l (5 gal.) batch of slip that has been insufficiently mixed and contains areas of high binder concentration will yield a tape with areas of high binder concentration. Similarly, a 20 l slip batch whose mean particle size increases from the top to the bottom of the storage container will tend to yield a tape with varying mean particle size along the length of the cast. Chapter 3 addresses the issue of obtaining a homogeneous slip. Great care taken to generate a uniform and homogeneous slip, however, does not ensure that the slip will remain homogeneous.

Many, if not most, of the slips made for tape casting are stable for a long enough period of time to be considered homogeneous during the casting process. The presence of a long-chain polymer in relatively high concentrations adds stability against flocculation of the inorganic particles, though this may be a problem in aqueous suspensions. The steric action of the dissolved binder also acts as a strong suspending agent for most particles over an adequate period of time. Some powders, however, do not follow these generalizations.

Perhaps the most common slip homogeneity problem faced in tape casting is the settling of suspended particles. While settling is not a major issue of concern for small, light particles, it can pose a major processing hurdle when casting large, high-density particles. While submicron alumina (ρ = 3.986 g/cc) powders are easily held in suspension for four hours or four days, 150-μm tungsten carbide (ρ = 15.6 g/cc) is difficult to keep suspended for four minutes. Both the particle size and the material density are factors in the suspension stability, since settling is a function of particle mass. High-density particles can be held in a stable suspension if the equivalent spherical particle diameter is very low. On the other hand, extremely large particles of a low-density material can be extremely difficult to suspend. This is also seen in poorly deagglomerated slips where each agglomerate of fine particles acts as a single particle with a large equivalent diameter. Experience has shown that the stability of a slip against settling is not a function of particle buoyancy, but of chemistry and viscosity. Again, the long-chain polymeric binder is the most effective suspending and stabilizing agent in the slip. Where powder settling is a concern, an increased binder content can sometimes add the desired stability and maintain a homogeneous slip. "Mechanical" stabilization can also be an effective tool; a decrease in solvent concentration can raise the slip viscosity to the point where settling is no longer observed.

Settling is most commonly observed in slips made with metal powders and/or blends of metal powders due to their relatively high density. Metal powders for tape casting also tend to have a relatively high particle size in comparison to ceramic powders to be cast, usually due to the final applications. Settling and the resulting segregation are also areas of concern in slips with multiple ceramic powders where powder densities differ significantly, as in the alumina/zirconia system. Ceramic/metal powder blends are perhaps the most difficult slips with which to maintain homogeneity due to the large differences in particle mass. When settling is a concern, we always recommend adjusting the particle size of each powder component so that all of the particles have the same mass. This may or may not be an option in light of further processing steps and/or final fired part performance.

Constant agitation and stirring is in order for all slips prior to casting, and it is usually mandatory for slips that have a tendency to settle. Some powders described by large particles with high densities will always exhibit a tendency to settle upon standing. When polymer overloading and high viscosities are inadequate to maintain slip homogeneity and good suspension of particles, the only remaining course of action is constant stirring of the slip, preferably a stirring action that imparts an upward motion of the particles. It is not by any means common, but also not unheard of, to continue stirring the slip to avoid settling even in the slip reservoir by the doctor blade.[15]

4.2.2 Viscosity

Viscosity is one of the four primary defining variables for tape thickness. The viscosity of a slip greatly affects the flow characteristics under the doctor blade. Viscosity is described as a definite resistance to change in form or "internal friction."[16] As mentioned previously, the doctor blade wipes off the excess slip deposited on the carrier surface. The result of this metering process depends partly on the viscosity (internal slip friction) of the fluid being metered (see Figure 4.19). Since there is a pool of slip, or "head," on one side of the doctor blade, and a much smaller amount of slip on the other, a net hydraulic force is created that compels the slip to flow under the doctor blade. This hydraulic force stems from the weight of the slip in the reservoir (head pressure) and affects the volume flow under the blade during the casting process. The viscosity of the slip determines the extent of fluid flow due to the forces acting on it. A low-viscosity slip will deform to a greater degree than a high-viscosity slip under the hydraulic pressure of the reservoir. The flow under the doctor blade occurs, in many if not most cases, at a faster rate than the carrier speed. This "pushing" under the doctor blade has also been shown in mathematical flow modeling for both Newtonian[17] and non-Newtonian[18] slips. This flow, referred to as welling, depends not only on the slip viscosity but also on the head pressure formed by the reservoir height, blade gap, carrier speed, and blade shape. It does happen that the wet film thickness an inch downstream from the doctor blade is thicker than the doctor blade gap. We have actually cast tapes in which the dry tape thickness was thicker than the blade gap!

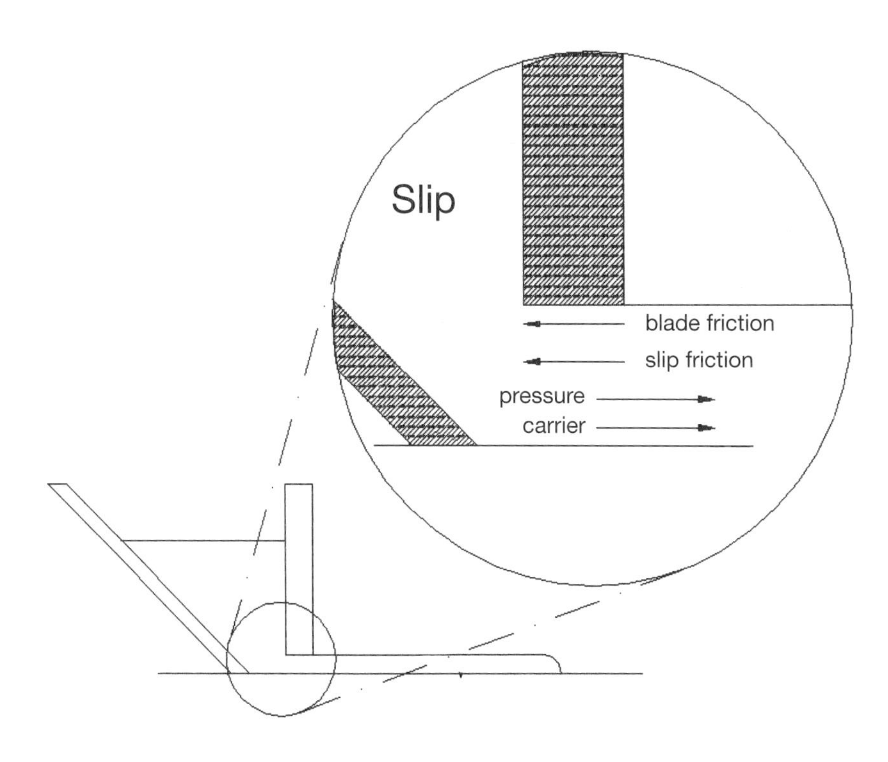

Fig. 4.19 Frictional forces and flow under a doctor blade during tape casting.

The welling action that takes place is directly affected by the viscosity of the slip, and it directly affects the thickness of the cast layer. The viscosity, therefore, becomes one of the four main determining variables of final tape thickness. The fact that viscosity has such a large effect on final tape thickness is the reason that so much care is taken during the slip preparation process to create a slip with repeatable viscosity from batch to batch. The viscosity of a tape casting slip is affected by many variables, all of which can be traced through viscosity into the final tape thickness. After slip preparation is complete, the biggest variable affecting viscosity is slip temperature.

Section 4.1 mentions control of slip temperature prior to casting. This temperature control is to provide a uniform viscosity from batch to batch in order to cast a repeatable tape thickness batch after batch. Some manufacturers control slip viscosity by controlling temperature, being less concerned with the actual temperature than with the resulting viscosity. The temperature effect on the slip stems from the fact that most tape casting binders are thermoplastic resins with low T_g (intentionally modified by the Type I plasticizer from Chapter 2). The other factors that affect slip viscosity, such as powder surface area, and slip preparation consistency, are much harder to control and are usually tolerated within a range of values. The issue of control over the viscosity becomes fairly clouded when we consider that slip viscosity changes with shear stress. That introduces the concept previously referred to as "fluid nature" or "rheology." It also reminds us of a popular quote, "Control is an illusion." Some manufacturers limit the effect of viscosity on tape thickness, thereby reducing variations, by increasing slip viscosity into the "paste" region (30,000 to 45,000 mPa·s [cP]). This is almost an extrusion process and, while it is done, it is not commonly found in industry.

4.2.3 Fluid Nature

Fluid nature has been called many things, such as rheological property, flow behavior, or the specific nature displayed by a slip. The practical implication is that the fluid being cast changes in viscosity with changing shear stress, and the casting process introduces the slip to many different shear forces. Not only that, but the extent to which shear affects viscosity is also variable with respect to time. Particle suspensions have been studied in some depth as to rheological properties. Reed[19] describes several different types of rheological behavior and describes some of the mechanisms that cause these differences. The vast majority of tape casting slips are best described as pseudoplastic.

Pseudoplastic slips are characterized as having a "shear-thinning" nature. The viscosity of a pseudoplastic tape casting slip depends on the shear stress applied at the time (including the shear imparted by the measuring device). For this reason, viscosities for tape casting slips should always be reported together with the shear conditions of the measurement. This behavior is beneficial in the tape casting process, since the slip will display a lower viscosity under the shear of the doctor blade, yet a higher viscosity downstream from the blade, thereby resisting motion within the metered film. This pseudoplasticity also allows thicker tapes to be cast without undue lateral spreading of the slip after blading. The pseudoplastic rheology of the slip also increases the usable width of the cast tape, since the edges of the tape will not taper out to as great a degree as would be seen with a Newtonian slip. Loss of control arises, however, since the viscosity is not constant from place to place in the casting process. Actual slip viscosity will differ in the holding tank, through the filter, in the reservoir, under the blade, and downstream from the blade. Referring back to the "welling" phenomenon, the viscosity (which significantly affects welling) is different in the reservoir head than it is when flowing under the doctor blade. In manufacturing, these differing viscosities are not controlled. Other more controllable variables, such as blade gap and reservoir head, are adjusted to account for small changes in viscosity and/or rheology.

Another rheological property often displayed by tape casting slips is thixotropy. Thixotropy describes the tendency of a slip to increase in viscosity over time under no shear. With no shear applied to a tape casting slip, a short-range order begins to form in the polymer network. The polymer chains form weak bonds throughout the slip, creating a network similar to a glass matrix. The network formed does not have long-range order as would be seen in a crystalline matrix, but instead would be more properly described as having the short-range order of a glass network. The time-thickening character of a tape casting slip has one benefit among the myriad drawbacks. The only benefit seen so far has been in the research area, where a trial formulation yielded a slip with far too low a viscosity to cast the desired thickness of tape. The slip, aged for four hours under no shear, increased in viscosity almost tenfold and was able to yield a very thick tape without spreading out after the blade. With that exception noted, thixotropy is considered a hurdle to be overcome.

The most common, and most commonsense, solution to thixotropy (thickening over time under no shear) is to store the slip under shear. The second-most-common solution to the thixotropy problem is to avoid storage time between discharging, de-airing, and casting. Aging of slips in the ready-to-cast condition is not recommended due to potential binder–dispersant interactions (see Section 2.6). Most practitioners in the tape casting field would likely recommend using both solutions to limit the extent of thixotropy. In manufacture, slip is usually discharged into a pressure tank fitted with an air-driven stirring paddle to agitate the slip. While the main purpose of the agitation is to impart shear to lower the slip viscosity to aid in degassing, the agitation also avoids most of the thixotropic effect. Slip should be cast as soon as is reasonably possible after degassing but, in a high-volume casting process, "as soon as possible" may be two days or more. A common solution to thixotropy is to store the slip in the holding tanks with the stirring paddle still in operation. The thixotropic effect on slip is also guarded against on its way to the doctor blade due to shear forces imparted by flow through feed tubing and especially by flow through a slip filter.

The combination of thixotropy and pseudoplasticity in a tape casting slip generally adds up to yield-pseudoplasticity. Yield-pseudoplasticity describes a slip that displays reversible deformation at the application of a shearing force up to a yield stress, at which point the weakly bonded (gelled) matrix begins to break apart, displaying a pseudoplastic rheology. The yield stress is sometimes referred to as "gel strength." In this case it should be noted that the change over time from pseudoplastic to yield-pseudoplastic is a reversible change, reversed by the application of high shear to break apart the binder network and return the slip to the "as-discharged" rheology.

Occasionally a fourth rheological effect will be seen in a highly loaded tape casting suspension. The high solids loading in tape casting slips will sometimes generate a behavior known as *dilatancy*. Dilatant fluids display "shear-thickening," whereby the viscosity increases with increasing shear. Dilatancy is never beneficial in the tape casting process and should be avoided. The simplest way to picture dilatancy of a highly loaded fluid is to envision so many particles suspended in the fluid that, when particles try to move within the fluid, they run into each other and lock together into a solid-like form. Dilatancy is avoided by lubricating the particle/particle interface. Dilatancy is most often seen with angular particles that can lock together easily. It is almost never seen with spherical particles, since the particles cannot lock together. Dilatancy can easily be avoided by adding more solvent to the slip, thereby increasing the space between suspended particles. The Type II plasticizer, as mentioned in Chapter 2, is also a solution to dilatancy, since it acts not only as a lubricant in the dry tape but also as a lubricant in the slip prior to casting. The lubricating nature of the Type II plasticizer provides a slippage layer between particles, resulting in good flow under shear. In our experience, dilatancy is a rare phenomenon in tape casting slips due to the relatively high organic content of the slips. It is most commonly seen in slips made with large, high density particles where solvent content has been decreased to avoid settling and where the binder content is low.

4.2.4. Surface Tension and Wetting at the Slip/Carrier Interface

These two topics are very closely related. The tape casting process spreads a thin layer of fluid onto a surface. The surface tension of the slip and its interaction with the surface energy of the carrier determine the behavior of the film after the blading process is finished. The wetting behavior of fluids has been discussed in great detail by a number of authors and is outside the scope of this book.[20,21,22,23] The effect of wetting (or de-wetting) on the tape-cast sheet, however, is an important factor in generating a good tape. Wetting of the carrier surface is necessary in order to maintain a thin, uniform layer of slip on the carrier surface prior to drying. The stability of the thin fluid layer is a balance between forces acting on the fluid. The weight of the fluid, the viscous and rheological resistance to motion, and the energy balance at the liquid/solid/air interface will determine the shape of the tape edges (see Figure 4.20). Energy balance will induce the edges of the wet film to taper at an angle (θ) defined by the interfacial energy balance. This contact angle at the tape edge will encroach to some extent into the center of the cast tape. Slip viscosity, specific gravity, and rheology can modify the actual contact angle and therefore the size of the tapered area. Due to this edge tapering, the width of uniform thickness will be less than the cast width. Organic solvents have a significantly lower surface tension (20 to 30 dynes/cm) than water (73 dynes/cm)[24] and do not normally display any significant interfacial changes or significant changes in edge tapering with changes in carrier material. Water, on the other hand, has a large contact angle in contact with low-energy surfaces.

Water is seen to dewet (physically flow across the carrier surface into beads) from low-surface-energy carriers. The incompatibility of water with low-surface-energy solids is well known to anyone who has ever waxed their car before a rainstorm. The contact angle of the water is often high enough that a 15.24-cm (6 in.) wide cast 0.0508 mm (0.002 in.) thick dewets to a 6.35-mm (0.25 in.) wide cast 1.27-mm (0.050 in.) thick. With high-water content slips, it has even been seen that the slip dewets from the carrier surface to so great an extent that the fluid never left the doctor blade reservoir as the carrier moved underneath.[25] The dewetting action is caused by an improper balance between carrier surface energy, slip surface tension, slip specific gravity, and slip viscosity/rheology. A higher surface energy carrier such as polypropylene will greatly lower the contact angle and be "wet" much more easily than a lower surface energy carrier such as silicone. In the same manner, lowering the surface tension (liquid/vapor interface) of the water with a "wetting agent" will also decrease the contact angle. Working from a different angle, increasing the viscosity, pseudoplasticity, or specific gravity of the slip will increase resistance to motion within the slip itself and limit the reaction of the fluid. We have overcome several dewetting problems in aqueous slips simply by increasing the solids loading in the slip, thereby increasing specific gravity, viscosity, and pseudoplasticity and avoiding the dewetting.

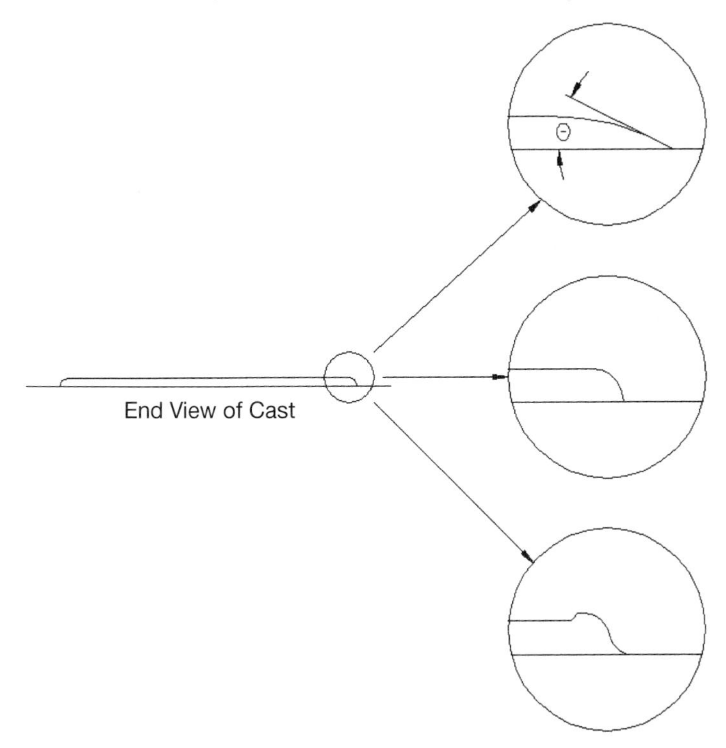

Fig. 4.20 Cast tape edge profile.

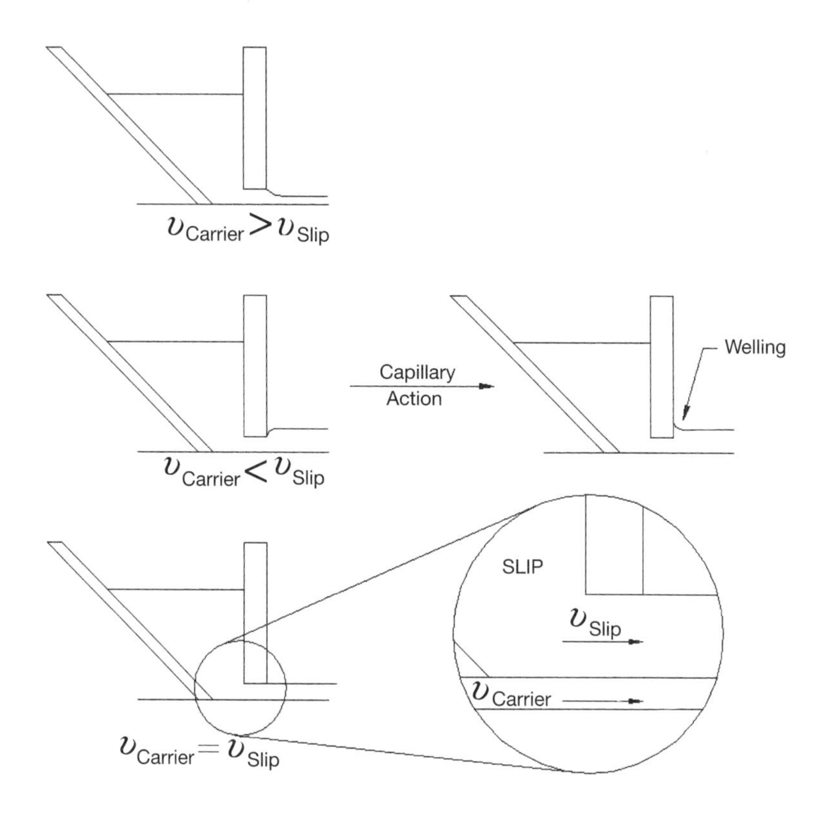

Fig. 4.21 The effect of carrier velocity on wet tape thickness.

4.2.5 Carrier Speed

Carrier speed, often called casting speed, determines the production rate of tape. Carrier speed is one of the four primary variables controlling tape thickness. The surface moving under the stationary blade and reservoir imparts a motive force, essentially dragging the slip out of the reservoir. It is interesting to note that, of all the forces at work in the blade area (see Figure 4.19), the carrier motion is the only externally applied force. The motion of the carrier creates a forward force on the fluid, compelling it to move under the blade and into the drying chamber of the machine. The motion of the carrier creates a shear force both in the reservoir and under the doctor blade, altering the viscosity of the pseudoplastic slip and thereby reducing the "slip friction" force. The flow volume under the doctor blade is partially governed by the slip viscosity and is therefore greater at higher carrier speeds. Increasing carrier speed not only adds more "carrier force" to the fluid flow equation, but it also decreases the slip resistance factor. This increased volume flow is not, however, in direct proportion to the increase in carrier speed, which means that, even though the volume flow is greater, it is spread over a greater area, yielding a thinner cast layer. Increasing carrier speed, all other variables held constant, will decrease green tape thickness. It is very important to hold carrier speed constant during a cast. Changes in carrier speed will cause direct changes in green tape thickness.

As can be seen in Figure 4.19, the force imparted by carrier motion works in concert with the head pressure of the reservoir, promoting volume flow under the blade. The volume of slip flows through a fixed area (orifice) defined by the carrier surface on the bottom, the doctor blade on top, and the side walls of the reservoir. Dividing the slip volume flow by the fixed orifice area yields a net slip velocity through the doctor blade opening. The wet thickness of the cast layer (wet tape thickness) downstream from the doctor blade is defined by the ratio of slip velocity to carrier velocity. As can be seen in Figure 4.21, the carrier speed has a significant effect on the wet thickness of the cast. It bears repeating that slip velocity is affected by viscosity which, in a pseudoplastic slip, is affected by the shear force imparted by the carrier and varies with carrier speed. Slip velocity is rarely, if ever, calculated in a manufacturing process, since carrier speed must be modified by other, more important factors. The main concern with carrier speed is the consistency of the carrier speed under the doctor blade.

Carrier speed consistency can be affected by a number of different factors. These include improper speed reference, drive chatter and motor variability (discussed in Section 4.1), and variable tension of the carrier film. Not all casting surfaces experience all of these variables. Variable tension is not usually an error source when casting on a metal belt or on a rigid substrate such as granite or glass. Polymer film carriers, however, require some attention to tensioning. These carrier speed variables were all discussed in Section 4.1.

Carrier speed in a continuous cast (cast, dried, and rolled in the same process) is usually defined by the required production rate. The casting speed is limited in a continuous cast by the drying rate of the cast layer, since the tape must be dry at the point where it is rolled or discharged from the casting machine. In a batch cast (cast, stopped, and allowed to dry before discharging from the machine), there is no limit to casting speed other than that imposed by the casting equipment. In batch casting, and in some continuous casting, the carrier speed is sometimes adjusted based upon the slip flow characteristics (for example, very low-viscosity slip may require increased carrier speed to avoid exaggerated welling). Since the extent of welling is not a controllable phenomenon, it is often beneficial to increase the blade gap and increase the carrier speed in order to obtain the same mean tape thickness with less welling.

Most production-scale tape casting equipment is supplied with a mechanism to control the linear carrier speed. The previous discussion of the carrier speed refers to the linear carrier speed in the region of the doctor blade assembly. Continuous casting machines that roll up the used carrier film onto the drive roller must be equipped with some mechanism to reduce the drive roll revolutions per minute as the diameter of the winding roll increases. With constant rpm and an increasing roll diameter, as carrier is wound on, the linear speed under the doctor blade will increase, causing a gradual decrease in mean tape thickness along the length of the tape. This was also discussed in Section 4.1.

4.2.6 Doctor Blade Height (Blade Gap)

The distance between the carrier surface and the bottom of the doctor blade is the most important variable in determining wet-film thickness. The other three primary variables (viscosity, reservoir height, and carrier speed) have less impact than the doctor blade height. The distance from carrier surface to blade bottom is usually referred to as blade gap. Interestingly, the blade gap does not directly affect tape thickness but instead defines the flow orifice that affects net slip velocity. Essentially, slip velocity works together with carrier speed to determine the base wet thickness value, which is then modified by the other factors such as drying shrinkage and dewetting. Constrictive flow through an orifice is a field unto itself and is outside the scope of this book. The inhibitive force of the orifice is described as "blade friction" in this book for conceptual use.

As seen in Figure 4.19, the blade gap determines the orifice through which the slip flows. Since cast width, reservoir width, and blade width are nominally identical, they cancel each other out of the slip flow equations, leaving blade height as the only variable determining the size of the orifice. All widths being equal, it is acceptable to view the flow in Figure 4.21 as an area flowing through a height at a given velocity (νslip). Changes in the height (blade gap) would cause proportional changes in the slip velocity and thus affect the slip height after casting. Changes in blade height, however, do not linearly affect wet thickness. Changes in the blade gap cause changes in the shear forces under the doctor blade. As we noted earlier, changing the shear force on a pseudoplastic tape casting slip modifies the slip viscosity (slip friction) and thus changes the volume flow of the slip. In general practice, changes in blade friction (Figure 4.19) have a larger effect on slip flow than do changes in slip friction (viscosity). While increasing the blade gap may decrease shear and thus increase slip friction, the reduction in blade friction has a larger effect, causing a higher wet thickness/blade gap ratio. In other words, increasing the blade gap by 10% may increase the wet thickness by 15%. The converse is also true. Decreasing the blade gap from 250 μm to 200 μm (0.010 in. to 0.008 in.), a 20% decrease, may decrease green tape thickness from 150 μm to 110 μm (0.006 in. to 0.0045 in.) a 25% decrease.

Because of its importance in determining tape thickness, great care must be taken to create and maintain a consistent and uniform blade gap. The doctor blade itself is usually surface ground and polished to obtain a smooth, flat surface. The doctor blade gap is then set using micrometers, one on each side, to create a uniform gap from side to side. The blade gap uniformity tolerance is typically a balance between tape thickness and cost. Thicker tapes will tend to have a greater inherent variation in thickness due to other factors, while thin tapes show a much greater thickness sensitivity to blade gap. The sensitivity of tape thickness to blade gap uniformity is balanced against the cost of creating and maintaining extremely high flatness tolerances. A local machine shop with a surface grinder can often grind a flatness of ± 5 μm (0.0002 in.) onto a stainless steel blade for $100 to $350. This is usually adequate for tapes over 50 μm dry thickness that are less than 25 cm wide. As casting width increases, the length of the stainless steel blade and the corresponding flexure of the blade can become a problem. For thinner and/or wider tapes, or extremely pseudoplastic slips, more rigid materials and tighter flatness tolerances may be necessary.

Special alloys with high flexure strength and high abrasion resistance have been employed in some tight-tolerance applications to ensure uniform blade gap. Over time, the blade surface will be abraded by the slip due to the high loading of powders. This abrasion makes the blade gap nonuniform and consequently makes the cast tape nonuniform. Blades should be remachined at regular intervals to ensure a flat, smooth surface. As mentioned, abrasion-resistant materials are commonly used to increase the working life of the blade. Some manufacturers make the doctor blade out of the same material being cast to resist wear. In some cases, the casting head itself (doctor blade or other metering assembly) can cost more than the rest of the casting machine.

The uniformity of the blade gap from cast to cast can also be a source of variation in tape thickness (wet and dry). Many manufacturers use an adjustable mounting system for the doctor blade, which is set to height before each cast. Many of these blades are set by hand and are subject to human error. This human factor can significantly affect repeatability. To avoid this, some manufacturers simply minimize the human factor by having the same person set the blade before each cast. This practice shows the casting procedure to be a "manufacturing art." Others choose instead to use a fixed-gap doctor blade (Figure 4.6) in which blade gaps are precisely machined into the doctor blade. The fixed-gap doctor blade then rides directly on the casting surface, giving a perfectly repeatable blade gap from cast to cast. Depending on blade style, slip viscosity, and the height of the gaps, up to eight different gaps can be machined into the fixed-gap assembly.

There are significant benefits and drawbacks to both the adjustable and fixed-gap doctor blade assemblies. The benefits and drawbacks of the different styles of blades are all tied to the ability to set the proper, uniform, accurate, and repeatable blade gap. While the fixed-blade assembly gives an accurate and repeatable orifice, slight changes in slip viscosity or rheology (normal manufacturing variations) may require slight corrective adjustments in blade gap that are not possible with the fixed-blade configuration. These corrective adjustments are possible with an adjustable doctor blade assembly. While the adjustable-blade configuration can be raised or lowered to account for normal variations in the slip, it does not lend itself as readily to accurate and repeatable initial settings. The adjustable doctor blade can also be used to cast a variety of different green tape thicknesses with a single assembly, whereas the fixed blade is limited in the number of different thickness tapes it can produce. It is very common to start with an adjustable doctor blade assembly to determine the proper blade gap over a number of casts and to then have a fixed blade made.

4.2.7 Doctor Blade Size/Shape

The shape of the doctor blade has been a subject for debate for almost as long as tape casting has existed. Figure 4.5 shows a number of different shapes that have been used in the field. Other shapes have been used in addition to those shown. The doctor blade shape used is influenced by a number of different factors, ranging from blade cost to maintenance cost to comfort. There are two features of doctor blade shape that have an influence on the quality and uniformity of the cast, the upstream corner and the downstream corner. The upstream corner can, in some cases, affect slip flow under the blade while the downstream corner comes into play when welling exists.

Grandma's Cooking Pot

A young mother was teaching her daughter how to cook a roast. She walked her through choosing a good roast and seasoning the meat. She then cut a one-inch slab from the end of the roast, put the remaining meat into the roasting pan and put it in the oven. The one-inch slab was cooked on the stove top in a deep frying pan. The daughter asked, "Why do we cut the end off the roast? Why not just roast it all?"

The girl's mother paused for a second, pondering the answer and replied, "That's how my mother taught me to cook a roast."

"But mom," the girl insisted, "How come?"

A phone call was made, and the question asked, "Mother, when you taught me to cook a roast, we cut an inch off the end and fried it on the range. Why did we do that?"

The woman's mother explained, "That's how I was taught to cook it, dear, do you remember how your father loved my roasts?"

The woman then called her grandmother, hoping to answer her daughter well. "Gram! It's your granddaughter. I'm in a fix. I'm teaching your great-granddaughter to cook a roast and she wants to know why we cut a slice from the end to fry on the range. I called Mother and she said that you taught her. Do you remember?"

"Oh! Yes I remember," her grandmother replied. "The butcher in town was always sweet to me. I think he liked me. He always saved the biggest and best roasts for me, trying to win my heart. I didn't have the heart to tell him that they didn't fit into my roasting pan. I always had to trim the roasts down to fit them into the roasting pan."

The first logical point to address is comfort.

Tape casting is not immune from the effects of technology transfer from other fields. The doctor-blading process is far from a new process, and it was used long before Glenn Howatt tried making thin ceramic sheets with it. Many coatings are applied using a doctor blade process, including paper and textiles. High-speed coating lines (200 to 300 feet per minute or faster) have shown that a bevel on the upstream corner (reservoir side) of the doctor blade creates a downward pressure on the slip and yields not only a better coating of the material, but a more uniform flow in the slip reservoir. This concept is still used in the tape casting field and can be seen in published figures on the tape casting process.[26, 27, 28, 29] One of the earliest published figures shows a schematic of a fully beveled blade[26] with no flat section at all (effective thickness close to zero). Some practitioners in the tape casting field still insist on the upstream bevel. Beveled blades are shown in Figure 4.5(c,d,g,h,l).

This lead edge bevel is not a detriment to the tape casting process, but neither is it always necessary. While it may be a distinct advantage for casting tapes at speeds over 10.7 m/min (35 ft./min), tapes over 0.254 mm (0.010 in.) dry thickness are usually cast in the 0.305 to 1.22 m/min (12 to 48 in./min) range or slower. The slower-speed casting processes negate the advantage of the beveled blade. We have seen no appreciable effect of doctor blade shape on tape quality or uniformity for casting speeds up to 4.6 m/min (15 ft./min). The bevel is an example of “grandma’s cooking pot” in the ceramic field. One thing that *is* accomplished by beveling the doctor blade is shortening the effective thickness of the doctor blade and thereby lowering the blade friction component of slip velocity.

Blade thickness, or “effective blade thickness,” is a large factor in the blade friction component of slip flow. The thickness of the blade determines the length of the orifice through which the slip must flow. Thicker blades extend the region of reduced area, thereby resisting slip flow to a greater degree and lowering slip velocity. Studies performed with various blade thicknesses similar to Figure 4.5 (a,b) have determined that net volume flow was reduced as blade thickness increased. Taken to extremes, the blade friction can be increased by increasing blade thickness to the point where it overcomes reservoir head pressure, making the blade gap the only significant variable in the wet-tape thickness. If this extreme is reached, it has been calculated and experimentally verified that carrier speed no longer affects tape thickness.[17, 18] This extreme case has not been reported outside of a laboratory. Some casting is done with extremely high-viscosity slips. Increasing the viscosity (Figure 4.19) to extremes accomplishes the same thing as increasing blade friction, making blade friction the only appreciable force on fluid motion under the blade.

Reducing doctor blade thickness decreases the blade friction component of slip flow, causing an increase in slip velocity and corresponding increases in the wet thickness of the tape. In a slow-speed casting process like tape casting, assuming 4.6 m/min or less, beveling of the doctor blade creates an effectively thinner doctor blade as shown in Figure 4.22. Without the intended benefit of the leading edge bevel (improved flow at high speeds), and with the potentially detrimental increased welling with a thinner effective blade thickness, and with the increased refinishing cost of a dual-surface doctor blade, it is recommended that a leading edge bevel not be used for tape casting. The bottom line is, "It doesn't hurt, but it doesn't help either. So why bother?"

This line of reasoning does not apply to the downstream corner of the doctor blade. As shown in Figure 4.23, experiments have been performed with different release angles on the downstream edge of the doctor blade. We have defined this angle as β, since we did not find it published in open literature. The downstream bevel of the doctor blade can cause serious problems, avoid serious problems, or be useless altogether, depending on other casting parameters being used. The problems surrounding the downstream bevel, which we will refer to as "release angle β" are associated with the contact angle θ of the slip in relation to the doctor blade material. This contact angle was discussed previously as it pertains to the slip/carrier interface. The slip has a similar interaction with the doctor blade. The tendency of slip to creep up surfaces is typically referred to as capillary action, and the results are shown in Figure 4.23. This capillary action creates a pocket of slip above the blade gap that experiences little to no flow.

The slip lingering on the downstream edge of the blade begins to dry and may, depending on the flow rate and drying rate, form semi-dry lumps of slip that can then "let go" of the blade, forming defects in the tape-cast sheet. Figure 4.23 shows the different release angles on doctor blades and the potential relationship to creating a low-flow zone after the blade. As can be seen in the figure, casting with a bevel on the downstream edge of the blade (back bevel) creates the greatest low-flow zone and will have the greatest potential for caus-

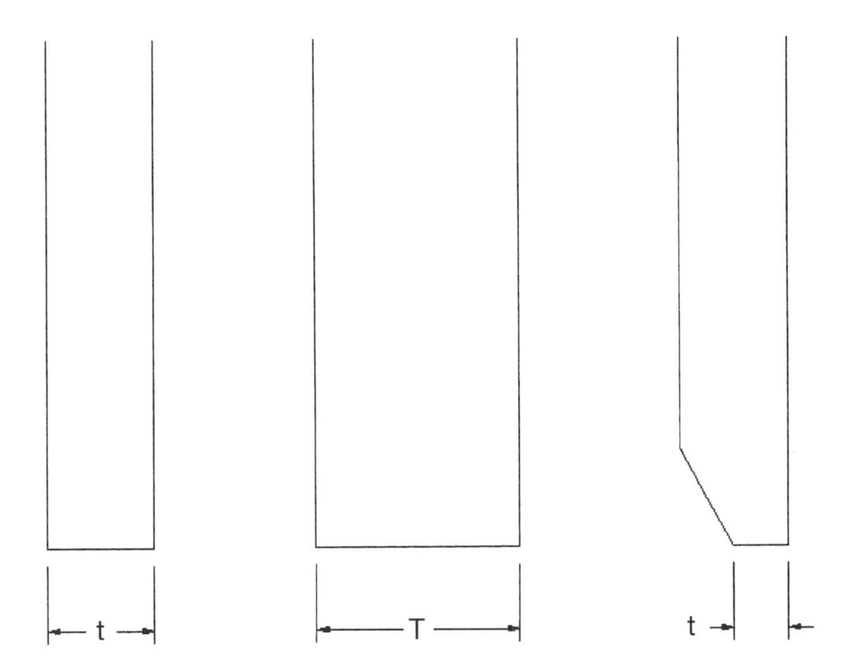

Fig. 4.22 Beveled and thicker doctor blades.

ing defects in the tape. We know of no argument for casting with a blade where $\beta < 90°$, whether flat or curved. Some practitioners cast with the back bevel anyway (Figure 4.5 [d,g]).

The low-flow zone can be mechanically avoided by increasing the release angle significantly beyond 90°, avoiding capillary action on the blade back entirely. Establishing $\beta > 90°$ eliminates the low-flow zone and thus promotes tape consistency and uniformity. A balance is typically drawn between dollars and danger, since a $\beta > 90°$ blade is more costly to manufacture and refinish than a $\beta = 90°$ blade. In

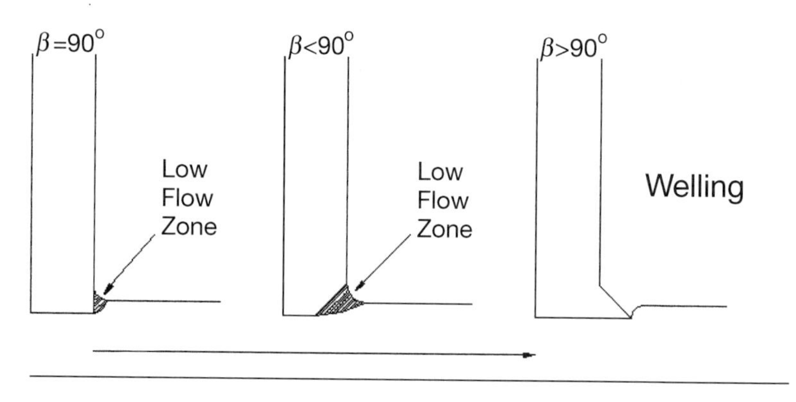

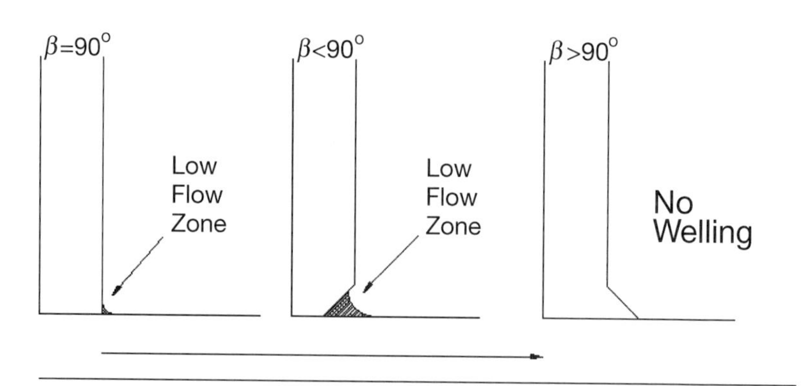

Fig. 4.23 Downstream doctor blade shapes.

practice, a $\beta = 90°$ blade is usually adequate if the slip is prepared and cast properly. While theory dictates $\beta > 90°$, it is usually not found to be worth the price.

4.2.8 Reservoir Height

Also referred to as "head" or "head pressure," the reservoir height is the fourth of the four primary variables controlling wet-tape thickness. Referring once again to Figure 4.19, the height of the slip determines the pressure pushing the slip under the doctor blade. Increasing the slip height in the reservoir increases the pressure, thereby increasing the slip velocity. As mentioned in the previous section, if the blade friction (thick blade) or slip friction (high viscosity) components are significantly greater than the pressure, the effect of changing the head height is minimized. In most casting processes, however, the blade is thin enough and the slip fluid enough to make head pressure a primary variable in tape thickness. As with the other variables influencing tape thickness, the head pressure must be tolerated within a certain range. The mean head pressure is not nearly as critical as the constancy of the head pressure.

While we refer to head pressure and reservoir height almost interchangeably, they are different and, in some casting setups, separable. The previous section on equipment described two types of casting, slot-die and lip coating, which do not have a reservoir height in the manner shown in Figure 4.19. In these two coating processes, the slip pressure is measured and controlled directly in units of pressure. The force balance in these types of coating is the same as is shown in Figure 4.19, but the pressure is externally applied. As mentioned at the beginning of this section, the bulk of discussion in this book is centered around doctor blading as shown in Figures 4.19 and 4.21 unless specifically mentioned otherwise. In the horizontal casting configuration shown in the figures, head pressure is a function of gravity acting on the slip and is therefore a function of slip mass. To mention again, the primary concern is not the amount of pressure but the constancy of that pressure.

The head pressure, or force, is defined by

$$\text{Force} = \text{mass} \times \text{acceleration} \quad \text{or} \quad \text{Pressure}_{slip} = \frac{(m_{slip} \times g)}{\text{orifice area}}$$

since the acceleration in this case is the acceleration of gravity (g). The slip mass m_{slip} is defined by

$$m_{slip} = \text{volume}_{slip} \times SG$$

where SG is the slip specific gravity (this should be nearly a constant). Slip volume is held in a reservoir where length and width are constant. The only variable left in these equations during casting is the slip height, which defines the volume, the mass, and the pressure. Since all of the other variables (SG, width, length, gravity) are constant, the pressure of the slip relies solely on the height of the slip in the fixed-area reservoir.

The previous section in this chapter discussed a variety of different control techniques for governing the slip height. Reservoir height does not usually have as large an effect on tape thickness as does the blade gap. In theory, this means that the doctor blade gap can be seen as a rough control for wet-tape thickness, while reservoir height can be used to fine-tune the wet thickness. Not many casters use this technique.

The optimum head height is a balance of other factors. Since the constancy of the head height is of great importance, the head height should be as high as possible so that the percentage variation in head pressure is as low as possible. With a low slip viscosity, slow casting speed, large blade gap, or thin blade, the head should be kept low to avoid excessive welling and the resulting thickness nonuniformity. A low reservoir head can lead to high variations in pressure (percent variation from average) with slight changes in head height. A high reservoir increases slip dwell time in the reservoir, which can result in skinning (premature drying of the surface slip) or settling in the reservoir. Balancing these two extremes will result in a usable, repeatable, and most importantly, constant head pressure for the casting process.

4.2.9 Fitting It All Together

There are four measurable and somewhat controllable variables that work together, either in complement or in opposition to each other, to determine the wet thickness of the cast film. These variables are called the four primary variables and are distinct and separate from each other: (1) Slip homogeneity, rheology, surface tension, and wetting are all factors that affect the viscosity (slip friction) under the casting blade. (2) Casting speed determines the externally applied force that compels the fluid to flow under the blade. (3) Blade gap and the effective thickness of the blade determine the flow orifice and thus the force of resistance applied by the doctor blade on the moving slip against the direction of flow (blade friction). (4) Pressure applied to the slip reservoir, whether applied by gravity or by external means, compels the slip to flow out of the reservoir under the doctor blade.

Any change in one of these variables can be offset by an opposite change in one or more of the other variables. A lower viscosity can be offset by a faster casting speed, a lower blade gap, or a lower reservoir head. A slower casting speed can be offset by a lower blade gap, a higher viscosity, or a lower reservoir head. Since these variables balance to create a given wet-tape thickness, a database is typically created to track cause and effect. In manufacturing, it is common practice to record viscosity, blade gap, and tape thickness for future reference. Since viscosity is tolerated within a certain range, it is often beneficial to modify the blade gap to account for batch-to-batch viscosity changes. In practice, it is most often the case that reservoir height and casting speed are held constant while blade gap is modified to account for changes in viscosity, thereby yielding a repeatable tape thickness.

This section has explored the variables that define wet-tape thickness. As the solvents are removed, the tape shrinks in the z-dimension so that the dry-tape thickness (green tape thickness) is less than the wet thickness. The following section explores the drying process.

4.3 DRYING

4.3.1 Introduction

Drying of the cast layer can be a whole field unto itself. With so many additives, and typically multiple solvents, drying of the tape and the behavior of the tape during the drying process can vary greatly from slip to slip. This section will explore the two major mechanisms that control drying rate along with some ways to modify them. The tape casting process—tape drying process actually—is somewhat unique among ceramic formation processes in that it is a one-side drying process. After the slip is metered into a thin and uniform layer, all of the solvent is removed from a single side of the cast. Two things work together to cause the one-side drying; a thin, essentially two-dimensional shape with no real height, and an impermeable carrier on the bottom. This single-side drying can be the cause of some very interesting activity within the tape matrix. Ideally, the chemical composition of the tape (primarily the solvent concentration) should stay uniform throughout the tape during the entire drying process. This, however, simply cannot occur, since all of the solvent must migrate to the top surface of the tape to evaporate. The two major mechanisms for drying in the tape-cast layer are those just mentioned: (1) rate of solvent evaporation from the surface of the cast and (2) rate of solvent diffusion through the tape to the drying surface. Of these two mechanisms, diffusion through the tape tends to be the rate-limiting factor.

The two drying components can be adjusted by various means. The volatility of the solvent at the tape surface can be adjusted by adjusting the types of solvent used, the concentration of solvent vapor in the local atmosphere, the local air temperature, and the solvent temperature. The diffusion rate through the body of the tape can be adjusted by altering the binder concentration, changing particle size, adjusting the wet film temperature, and keeping an open pathway to the surface. Some of these control techniques, such as particle size and binder content, need to be addressed during the preparation of the slip and factored into the initial slip recipe. Other parameters like air temperature, slip temperature, and local vapor concentration are controlled by the drying equipment separate from the casting slip. This section will take an in-depth look into the drying mechanism and some cause-and-effect relationships that can exist in the typical drying process. There are always exceptions.

4.3.2 Surface Evaporation

The means of escape for the solvent on the top surface of the cast layer is to take energy from the air and from the rest of the slip to evaporate into the surrounding atmosphere. The rate of evaporation is governed by the energy available to the solvent, the volatility of the solvent species, the vapor concentration of the local atmosphere, and the saturation concentration of the local atmosphere. Saturation concentration depends upon the gases in the atmosphere, the solvent species, and the temperature. Looking back into Chapter 2, if two or three solvents are used, they may form an azeotropic mixture and evaporate as a single solvent.

Since evaporation requires an input of energy, raising the temperature of the solvent will speed the surface evaporation process by providing an excess of energy. Raising the air temperature will not only provide the energy to evaporate (heat of vaporization), but will also increase the saturation concentration of the atmosphere, thereby increasing the distance to equilibrium. Air heating greatly speeds the surface evaporation rate (picture a puddle drying faster on a hot, sunny day compared with a cool autumn evening). Many tape casting machines are equipped with an air heating option to speed the surface evaporation of the tape. The saturation concentration of the local atmosphere will also affect the surface drying rate since, as mentioned, evaporation rate is affected by the system's proximity to equilibrium.

All things held constant, a liquid will evaporate more rapidly in an atmosphere with a very low concentration of the liquid's vapor than in a near-saturated atmosphere. For example, a puddle will evaporate more slowly on a very humid day than on a dry day due to the concentration of water vapor in the surrounding atmosphere (humidity). Due to this saturation effect, some practitioners opt to slow the surface drying rate of their cast by placing solvent-soaked rags or paper towels or containers of solvent in the drying chamber to saturate the local atmosphere. While effective, this practice is not advisable due to fire and safety concerns; it would send your safety engineer and local OSHA representative into fits.

Saturation or proximity to saturation is generally not an addressable concern with tape casting equipment due to the high levels of airflow. Airflow must be adjusted, apart from drying concerns, to a level specified by fire and safety codes as mentioned in the equipment section of this chapter. Minimum airflow calculations are shown in Appendix 2. For reasons to be described, measures are often taken to slow the surface drying rate as much as possible by reducing machine air flow to the lowest allowable limit. Additions of heat, differing solvent volatilities, and local concentrations of vapors can be controlled to affect the evaporation rate of solvents from the tape surface. The desired surface evaporation rate, however, is usually defined by the diffusion rate of solvent through the body of the tape.

Diffusion, or motion, of the solvent to the top surface of the tape is normally the rate limiting factor in drying. The rate of evaporation of surface solvent is normally so much faster than the solvent motion to the surface that a drying crust or "skin" forms across the surface of the tape. Efforts to limit surface evaporation stem from the desire to avoid this skin on the surface. Ideally, the solvent concentration should stay nearly uniform throughout the tape during drying so that all parts of the tape dry at the same rate. This would be accomplished by making the rate of diffusion equal to the rate of evaporation. The ideal case, however, is unattainable. In practice, the drying conditions, tape structure, tape components, and solvent mixtures are balanced to get as close to ideal conditions as the down-stream manufacturing needs allow.

Figure 4.24 graphically shows some hypothetical progressions of solvent concentration during the drying process. On the figure, *0* represents the carrier surface, *t* represents tape thickness (the drying surface), the uppermost graph represents wet tape, and drying progresses from the top plot to the bottom. As mentioned, the ideal case shows a uniform solvent concentration through the thickness of the tape throughout the drying process. In reality, the evaporation rate from the surface will always be much faster than the motion of solvent to the surface. Thus the best-case scenario displays a dry film on the top of the tape, yet a diminishing amount of solvent at the slip/carrier interface. This best-case scenario exists when the rates of diffusion and evaporation are as close to equal as possible. The worst-case scenario is realized when the rate of evaporation is much, much greater than the diffusion rate through the tape matrix. The top surface of the tape, giving off solvent much more quickly than the diffusion mechanism can replace it at the surface, forms an ever-thickening dry layer, while the solvent concentration at the slip/carrier interface does not significantly decrease. This is akin to the case-hardening problem sometimes seen in the spray-drying process. Due to the minimum airflow requirements imposed by safety and fire regulations, the surface evaporation can only be decreased to a certain extent. Due to this limitation, the other variables affecting surface evaporation require attention.

4.3.3 Solvent Diffusion

The speed at which the solvent can move to the surface is always the slowest function of drying. The rate of motion through the body of the tape is limited mainly by the body itself. The pathway for the solvent through the tape matrix is crowded with particles, binder, Type II plasticizer, and dispersant. As the drying process progresses, the tape shrinks, creating the dense, packed bed of particles that is the goal of tape casting. This dense, packed bed, however, limits the escape paths for the solvent at the slip/carrier interface. As binder concentration increases, the space between the particles becomes more and more filled with polymer, limiting the diffusion rate from bottom to top. Also, as particle size decreases or as PSD allows better packing density, the escape paths (diffusion paths) between particles shrink, thereby slowing the diffusion rate. Perhaps the biggest obstacle to diffusion through the matrix, however, is the skin.

As a rule, liquids diffuse much more quickly through a liquid medium than through any other medium. At some point in the drying

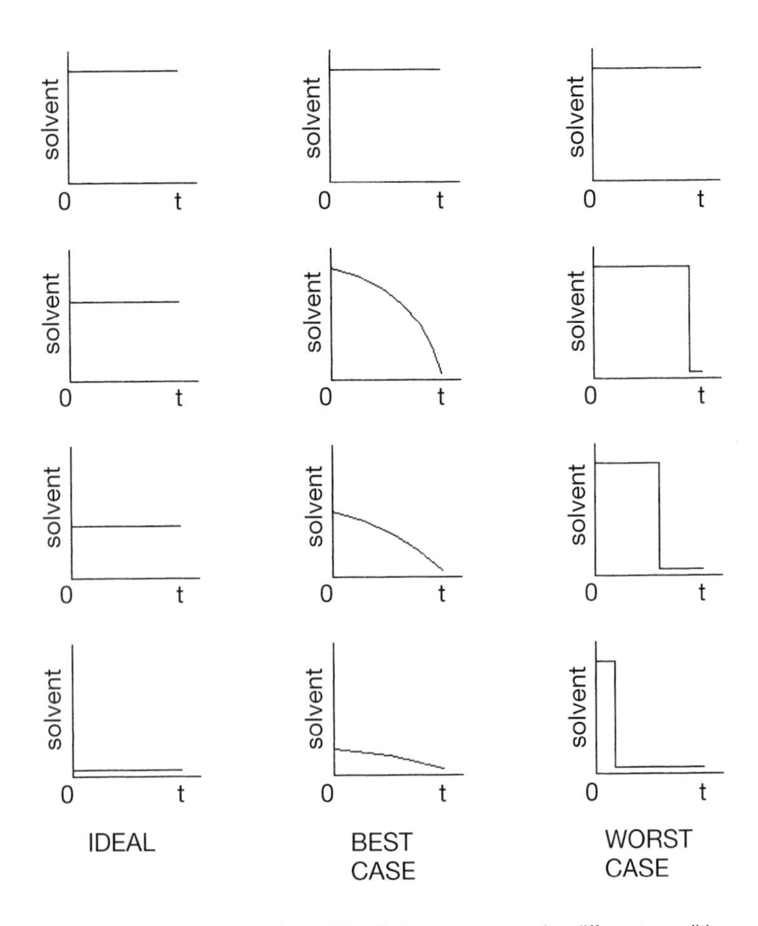

Fig. 4.24 Schematic representation of the drying process under different conditions.

process, the binder at the top surface of the cast will lose enough solvent to form a solid sheet or skin across the top of the tape. This is unavoidable since the dry tape is simply a solid piece of this skin. The diffusion rate of underlying solvent is much slower through this skin than through the liquid matrix of the slip. This, once again, is why effort is made to slow the surface evaporation rate . . . to delay the formation of this low-diffusion-rate skin. The dried polymer effectively plugs up the interparticulate spaces, creating a low-permeability layer across the top of the tape and limiting the bulk drying rate. This is where a balance must sometimes be drawn between drying rate and tape porosity. Allowing some porosity in the tape will increase solvent diffusion to the top surface by keeping an open pathway to the top surface. The addition of a slow-drying solvent can also aid solvent migration speed by delaying skin formation and providing a liquid pathway from bottom to top. This type of additive would properly be called a skin retarder and may actually be used as one of the primary solvents in the slip.

The last general variable affects both drying mechanisms. Heating the tape body itself not only increases solvent evaporation rate by heating the solvent, but also increases the diffusion speed of solvent through the matrix. Remember from the Binders and Plasticizers sections of Chapter 2, that the polymer matrix shares many of its characteristics with glasses. Having a high concentration of solvent effectively lowers the T_g of the polymer well below room temperature, causing it to behave as a liquid. As drying progresses, the solvent concentration decreases, increasing the effective T_g of the polymer. Heating the tape matrix either by air, radiant, or underbed heat raises the temperature of the polymer further above its T_g, promoting liquid-like behavior and increasing the diffusion rate of solvent through the binder. Many practitioners in the field find that the fastest way to dry a tape is to heat the bottom of the tape without heating the air. Heating the bottom of the tape increases solvent mobility in the tape body, driving the solvent up to the surface, while air heating tends to have a greater impact on the tape surface evaporation. In most cases, the surface evaporation doesn't need help.

Some "Cause-and-Effect" Relationships

One of two additional variables that was quickly glossed over earlier is the fact that the dimensions and structure of the tape are changing throughout the drying process. In the previous section on fluid metering, care was taken to refer to the cast thickness as the "wet thickness." This care was taken because the final thickness, the "dry" or "green" thickness, will always be less than the "wet" or "cast" thickness. This drying shrinkage is a fairly commonsense occurrence, since most items in everyday life shrink during drying. In the slip, the solvent accounts for a large percentage of the volume. When the solvent is removed from the system, the system is reduced in volume. This drying shrinkage, in most cases, is unidirectional in the z-direction (thickness). Motion in the tape matrix, therefore, is an unavoidable result of drying as the particles settle under gravity, the polymer chains shrink and reorient, and the tape structure increases in density. The tape, being cast onto a carrier that does not shrink, is held (bound) in the lateral directions, forcing all shrinkage to occur in the z-direction, inasmuch as the carrier does not change. There are exceptions to this rule, but through the majority of casting, the carrier remains flat and dimensionally stable. Exceptions to this are addressed later in this section.

Drying shrinkage (z-dimension) is a variable that the field knows little about outside of empirical observation and logical deduction. Shrinkage factors have been seen in our own experimentation to range from below 1 to approaching 6. Often referred to as wet/dry ratio, drying shrinkage is probably more accurately quantified as gap/dry ratio, since most practitioners do not measure the wet thickness of the cast. There are ways to measure the wet thickness, and some do, but the measuring equipment is typically more expensive than manufacturers are willing to pay. As mentioned in the previous section, we have cast some thick tapes (approx. 0.060 in. or 1.5 mm) whose dry thickness exceeded the blade gap due to welling. In these tapes, the gap/dry ratio is less than 1. (It should be mentioned that these tapes were cast to be aesthetically pleasing and had no thickness tolerance to meet.) In normal casting, gap/dry ratios typically fall within the range of 1.5 to 3.0. Gap/dry ratios as high as 8:1 have been reported for some systems.[30] Much like the slip viscosity mentioned in the previous section, it does not matter very much what the gap/dry ratio is, as long as it stays constant. With a constant gap/dry ratio, the dry-tape thickness can be adjusted simply by adjusting the gap as long as the adjustments are small.

Logical relationships exist, all of which we have empirically confirmed, between gap/dry ratio and other processing variables. Increasing casting speed increases gap/dry ratio (thinner tape), as shown in Figure 4.21. Increasing reservoir height lowers gap/dry ratio (thicker tape). Some changes cause offsetting results, such as increasing solvent levels in the slip, which increases the shrinkage ratio but also lowers viscosity and may increase wet thickness through welling. Increasing blade gap decreases the gap/dry ratio by lowering the "blade friction" (Figure 4.19) and increasing welling.

Blade gap is by far the largest determining factor in green thickness. The blade gap to dry thickness ratio, however, is not constant. Because of welling, increases in blade gap are not always in direct proportion to the thickness effect on the green tape. It would be possible to plot the gap versus green thickness relationship for a given slip, showing a slight curvature in the plot. As we mentioned, dry tape thicknesses can be adjusted simply by adjusting the blade gap, as long as the adjustments are small. For small increases in blade gap (a few percent or so) the relationship can be treated as linear, but for larger changes the curvature of the graph cannot be extrapolated as linear. Unfortunately, the degree of curvature is slip dependent and is affected by such variables as casting speed, viscosity, reservoir head pressure, wet thickness/dry thickness ratio, and solids loading. Essentially, the dry thickness of the tape is affected by a very large number of interdependent variables, not all of which can be controlled. These variables are instead balanced together with the blade gap in more or less a "trial-and-error" process to yield a green thickness.

Previously we mentioned that the drying of the cast layer causes shrinkage that is forced to occur in only the *z*-direction, through the thickness of the cast, due to the fact that the carrier film does not shrink and the cast layer is bound to the carrier surface. This relationship holds true for the cast layer since it is essentially a two-dimensional structure. In a two-dimensional structure, top and bottom are identical; thus when the bottom is bound by the carrier, the top is as well. As tape thickness increases to approximately three mils (75 μm) and thicker, the top and bottom of the tape behave as two different layers. The bottom of a thicker tape is still bound by the carrier, but the top surface is no longer bound due to the increased distance from the carrier surface. While a thinner (about 50 μm or less) tape will normally behave like a single layer, a thicker cast introduces some behaviors consistent with a three-dimensional structure.

The three-dimensional structure contains a top surface that is not bound laterally by the carrier film and a bottom layer that is bound laterally. Drying shrinkage for the top surface, and at least the top half of the body, is not forced to be in the *z*-direction only. The top surface of the cast, during solvent removal and the resulting shrinkage, sees lateral drying stresses in both the *x* and *y* directions (Figure 4.25). Any lateral shrinkage that occurs due to the evaporation of solvent will cause the top surface of the tape to be dimensionally smaller than the bottom surface that is bound. In the direction of casting, referred to as "cast direction," the carrier film is in motion and under tension; thus the only way for the top surface to be shorter than the bottom surface is for the tape to crack. Across the cast, referred to as the "cross-cast" direction (*y*-direction), however, the carrier is not always in tension.

The carrier film in the cross-cast direction is generally very narrow in comparison to the cast direction. The narrow width and lack of cross-cast restraint lower the force necessary to lift the carrier edges. Thicker tapes with significant lateral shrinkage will often lift the carrier edges, curling the edges of the cast and allowing the top surface of the green tape to be narrower than the carrier interface. Some machines are designed with a curved bed, convex in the casting direction, which prohibits edge curling of the carrier. On a flat bed, the extent of edge curling is determined not only by the lateral shrinkage tensile force, but also by the weight of the carrier, the weight of the tape itself, and other stress-relief mechanisms in place. Curling always begins with the tape edges, where the reduced thickness of the tape does not exert as much gravitational force. One of the other stress-relief mechanisms was covered in previous sections; the Type II plasticizer. The Type II plasticizer in a tape structure can act as a lubricant, facilitating motion within the tape structure in response to drying stress instead of curling.

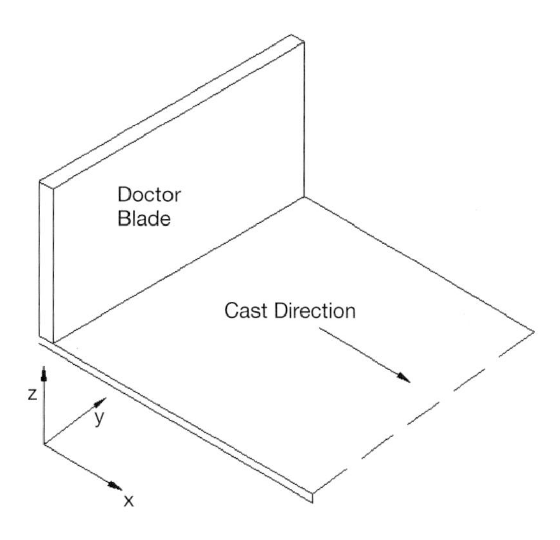

Fig. 4.25 Schematic of tape axes in respect to the doctor blade and casting direction.

Motion of the tape structure is necessary during the drying process for thicker tapes. At some point in the drying process, a fixed structure will be formed. As solvent leaves the system, the cast layer will slowly transition from a liquid solution to a solid sheet. This transition is gradual, often very similar to the cooling of a glass. While a glass becomes solid when its temperature decreases through T_g, the removal of solvent raises the effective T_g of the cast layer through the temperature. The "solid" tape still contains a fair quantity of solvent within the fixed matrix, but the matrix itself is defined during this liquid–solid transition. The fixed, wet matrix is sometimes referred to as being "gelled." Since the gel structure is formed while solvent is still in the tape, the structure itself is compelled to shrink as drying continues and solvent is further removed from the tape. The Type II plasticizer can often facilitate the continued shrinkage by promoting permanent "plastic" deformation within the matrix.

Literature surveys, field consultation, and troubleshooting, along with a great deal of proof testing, empirical observation, and hypothesis testing, have led us to believe the following:

The lateral stresses developed during solvent removal can cause a number of different effects, depending on other factors in the tape system. Repeating the Type II plasticizer effect, the internal structure of the tape body can permanently deform (yield) in response to the lateral stress. Also from above, the lateral stress can cause edge curling due to cross-cast shrinkage of the top layers. Regardless of the surface shrinkage allowed by curling, some of the tensile stress will be stored in the matrix and will remain in the tape after casting. The amount of stress that can be stored in the polymer matrix depends on the polymer type, polymer content, and effect of the Type I plasticizer. The Type I plasticizer increases the reversible strain in the tape matrix, which affects how much energy the matrix can store.

If the extent of lateral shrinkage is larger than can be accommodated by the combination of stored energy in the matrix, edge curling, and plastic deformation, the stress will be alleviated in one of three ways. If lateral shrinkage exceeds what can be tolerated, and if adhesion to the carrier surface is less than the particle-to-particle adhesion forces, the tape will disengage itself from the carrier surface, essentially freeing the bottom from constraint and allowing the bottom to shrink. If lateral shrinkage exceeds what can be tolerated, but the adhesion to the carrier is higher than interparticulate adhesion, the tape will crack and separate, thus relieving the tensile stress. If both the interparticle adhesion and adhesion to the carrier are high, the tape will simply continue to curl across the body of the cast. We have seen some dry tapes that curled to the extent that they looked like ancient scrolls and others that looked like baseball bats. Figure 4.26 shows the order of manifestation of the lateral stress-relief mechanisms.

Disengagement from the carrier surface is typically called "self-release" and is always preceded by edge curling. From Figure 4.26, any lateral shrinkage stress will cause various effects on the tape, starting with Level 1 and proceeding downward through the levels. The amount of stress varies with solvent content, the solvent content remaining after gellation, the quantity of binder, and other factors. It is often seen that a tape will only show Level 2 effects, or perhaps a minor amount of Level 3 effects. As the amount of lateral shrinkage increases, the tape progresses further across and/or down the tree. A tape can display evidence of more than one relief mechanism from a single level, but it will never skip a level. The order in which the stress-relief mechanisms of a single level manifest themselves can vary with various slip formulations and powder sizes. It is not only conceivable, but we have seen, that tapes can self-release and edge curl only, crack along the center and self-release, crack and curl through the body but remain on the carrier, and many other combinations. The order in which the shrinkage stress relief mechanisms occur is explainable after the fact, but at this point it is not fully predictable. As a general rule, the drying stress-relief mechanisms increase in number and in magnitude as green tape thickness increases. This is logical, since the stresses stem from differing shrinkage from bottom to top, and they would increase as the distance from bottom to top increases. These effects also increase in number and in magnitude as total solvent content increases.

Combining the dual drying mechanisms, evaporation and migration, with the shrinkage behaviors just described yields a plethora of possible drying behaviors. We will try to cover most of the behaviors and trace through the mechanisms causing them. Many different drying characteristics are regularly seen during tape drying, all of which are combinations of the individual drying behaviors listed in Figure 4.26. Some of the drying characteristics, like edge curling and stored tensile stress in the matrix, simply require allowances in the downstream processing of the tape and do not always mandate formulation or drying changes. Other drying behaviors like cracking are failures that prohibit further processing and require changes in the tape fabrication process. When casting thinner tapes (around 50 μm), the lateral stress developed during drying is minimal and rarely progresses into Level 3. Thicker tapes, as previously mentioned, can progress further down and across the drying tree. With the thicker tapes, especially those over 635 μm (0.025 in.), it is normal to experience multiple Level 4 mechanisms unless the ability to store stress in the matrix has been increased (Type I plasticizer) and plastic deformation has been promoted (Type II plasticizer). These thicker tapes have at least two different layers (top and bottom) and can display different drying behavior in the layers. Level 5 drying behavior only stems from slip inhomogeneity.

4.3.4 Flat Tape

Flat tape, whether forced flat by mechanical constraints or held flat by gravity or good chemistry, experiences Level 2 stress and may or may not undergo Level 3 "plastic deformation." We have never seen a tape that did not store residual tensile stress in the polymer matrix. This stored stress will be discussed in detail in later sections, along with methods for alleviating the stress. Plastic deformation cannot really be measured, but can be assumed when the same slip without a Type II plasticizer shows multiple Level 4 behaviors, but with the Type II plasticizer lays flat.

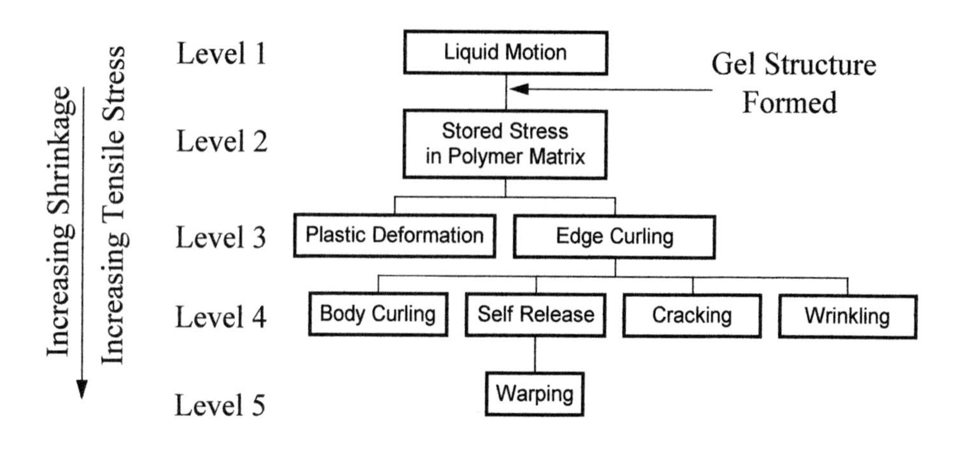

Fig. 4.26 Lateral stress relief mechanisms as a function of degree of drying in a tape.

4.3.5 Edge Curling

From Figure 4.26, edge curling is ranked equal to plastic deformation. Without mechanical restraint of the carrier edges, either by edge hold-down or curving the bed, most tapes exhibit some degree of edge curling (Figure 4.27). As discussed previously, edge curling is the result of the unbound top layer shrinking more than the bound carrier interface layer. The edges curl first, since the tape is thinner on the edges and therefore weighs less and lifts more easily. Slight edge curl is sometimes ignored in the manufacturing process since the thin edges of the tape are outside the thickness tolerance range and are trimmed off, anyway. Excessive edge curling can be a hindrance to the casting process and can be addressed by adding a small amount of Type II plasticizer, increasing Type I plasticizer, or heating the airflow, thereby softening the tape and promoting plastic deformation. Air heating may or may not be beneficial to combat edge curling, since it changes the entire drying profile. Air heating can cause more problems than it solves. Edge curling can also be combated by the mechanical restraint mentioned in this paragraph and in the "Equipment" section of this chapter.

Edge-curled tape, according to the drying tree, has already reached its maximum in terms of stored stress in the polymer matrix. The curled tape may or may not have undergone plastic deformation. The recommended Type II plasticizer addition is an attempt to promote plastic deformation instead of edge curling. Due to the flexible nature of dry green tape, this tape can usually be rolled for storage, which flattens the tape across the cast. It is often found that, after unrolling the tape, the edge curl is not noticeable.

4.3.6 Full Body Curl

Full body curl refers to the behavior sometimes seen during tape drying where the unbound top layer shrinks more than slight curling of the edges can account for (Figure 4.27). Full body curl requires three conditions to exist: high cross-cast shrinkage stress on the top layer, strong particle-to-particle binding force, and strong binder-to-carrier adhesion force. When the Level 3 stress-relief mechanisms are inadequate to alleviate lateral drying stress, the same mechanism that causes edge curling continues to act, curling the body of the tape. The difference between edge curling and full body curling is somewhat vague, since the cast will curl farther and farther in from the tape edge until an equilibrium is reached. We have seen very symmetrical U shapes in dry tapes as well as tapes that resembled ancient scrolls or baseball bats. This behavior is represented on the drying tree as progression through Levels 1, 2, 3, and into Level 4. Like edge curling, full body curling is sometimes tolerated in manufacturing, since the tapes can be flattened later in the process. The solutions available for full body curling (assuming constant powders) include: Type II plasticizer, heated air to soften the polymer, slowing solvent evaporation from the surface and thereby prolonging Level 1 liquid motion of the top layer, and radiant top heating to soften the top surface of the tape, which promotes plastic deformation of the top surface. When possible, we try to employ the Type II plasticizer, and we have seen tremendous success with this method.

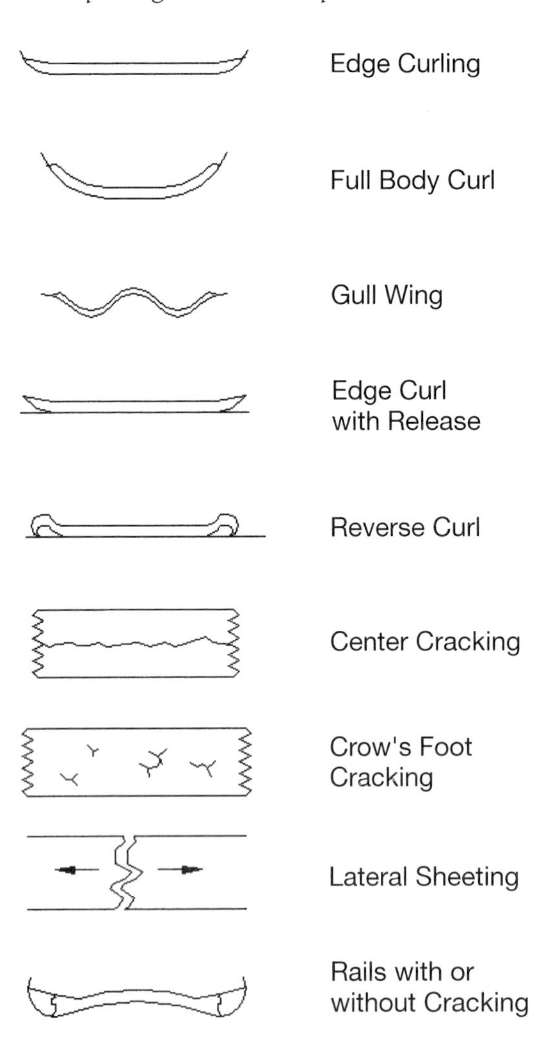

Fig. 4.27 Variety of curling and cracking observed in dry tapes.

4.3.7 Gull Wing

A variation of full body curl, gull wing curling stems from lateral shrinkage of the entire structure through the thickness of the tape. In order for both the top and bottom surfaces to undergo cross-cast shrinkage, the tape curves both up and down, ending up in the gull wing shape to balance top and bottom lateral shrinkage (Figure 4.27). The same conditions cause gull wing curling as cause full body curl: high lateral shrinkage of the top layer, strong particle adhesion and strong adhesion to carrier, with the addition of high lateral shrinkage of the carrier interface layer.

4.3.8 Edge Curl with Self-Release

If all of the drying stress components exist for gull wing curling (top surface shrinkage, bottom surface shrinkage, strong particle adhesion) but adhesion to the carrier surface is weak, the tape will progress down the drying tree through Level 3 edge curling to Level 4 body curling, then release the bottom surface of the cast from the carrier, allowing the bottom surface to shrink (Figure 4.27). Watching this take place, we have noted that body curling begins prior to the self-release. Upon further shrinkage of the bottom surface, the body curling is effectively undone to yield a flat tape with curled edges. We have never seen this behavior to undo the edge curl resulting in a flat tape. Edge curl occurs in Level 3 of the drying tree and is not undone by Level 4 effects.

4.3.9 Reverse Curling

During drying of a thicker tape, a solvent-depleted zone forms across the exposed surface of the tape. This solvent-depleted thickness gels or "freezes" into shape and size before the lower layers of the cast. Like a glass melt, the speed of transition affects the volume of the matrix. Faster cooling of a glass will yield a larger, less dense glass. Similarly, the fast drying of the top surface will occasionally freeze the top surface structure into a larger, less dense layer. In some cases, the bottom layers of the cast will shrink more than the top layers due to the slower drying rate (diffusion rather than evaporation) and force the tape edges, which had initially curled upward, to curl downward as drying is completed, lifting the tape away from the carrier (see Figure 4.27). This "reverse curling" is fairly rare, since it occurs primarily with thicker tapes and requires the high-mass body of the tape to be lifted upward.

Reverse curling is seen with very fine particles (which slow diffusion) in thick, high-density tapes. The additive effects of tight packing, very small particles, and high thickness create a large gradient of drying speed from top to bottom that can lead to the reverse curling. Reverse curling is not always considered a prohibitive defect and, while not optimal, it is sometimes tolerated. Like body curling, it can be avoided by the addition of a Type II plasticizer to promote plastic deformation instead of dimensional changes. Reverse curling can sometimes be avoided by heating the tape during drying, softening the polymer with heat and essentially annealing the stresses out of the matrix. Heating to avoid reverse curl, however, must come from underneath. Heating the top surface will speed the evaporation rate that is the cause of reverse curling in the first place. Heating the casting bed may or may not be effective to avoid reverse curl since, once started, the reverse curling of the tape lifts the tape away from the heat. Reverse curling requires high particle-to-particle adhesion, high thickness and packing density to slow the diffusion rate, and a weak adhesion to the carrier surface.

4.3.10 Center Cracking

All of the previous drying behaviors have had one thing in common: high particle-to-particle adhesion. When Level 3 stress-relief mechanisms are not adequate to compensate for the drying stress, and when interparticulate adhesion is less than carrier adhesion, the tape will crack and separate to permit lateral shrinkage. The most common cracking found in a uniform cast layer is cracking down the center of the tape (Figure 4.27). This essentially doubles the number of edges by creating two tapes, each approximately half the width of the original cast.

Center cracking does not always occur in the center of the tape. If the cast layer is nonuniform for some reason (streaks or furrows from metering, bubbles, agglomerates, etc.) the cracks will tend to nucleate at the nonuniformities and propagate along the length of the tape. If a significant streak (line of thinner material) exists along the cast, the crack will generally propagate along this defect. If a significant population of bubbles and/or agglomerates exists, multiple cracks often nucleate and propagate until the cracks connect with one another.

Center cracking can be avoided in a variety of ways. The first solution to try includes identifying the source of cracking. If the crack(s) all nucleate around a point defect such as a streak, bubble or agglomerate, filtering and proper casting to avoid streaks may solve the cracking problem. If this is not adequate, or if the cracks propagate through defect-free sections of the tape, other avenues must be sought. There is no such thing as a truly defect-free tape, and thus undue propagation must be addressed.

Slowing the drying rate by lowering airflow can extend Level 1 liquid motion and allow more solvent to leave the matrix prior to gellation. Additional Type I plasticizer can increase Level 2 stress storage, although this is typically not a large enough change to avoid cracking. Increasing Type I plasticizer may reduce Level 3 effects, but Level 4 effects are generally bigger problems than the Type I plasticizer can handle. Increasing Type II plasticizer can be a great aid to avoid cracking by promoting plastic deformation. The most effective solution to this problem, however, is to increase binder content. Cracking occurs when carrier adhesion is stronger than interparticle adhesion. Increasing binder content increases the free binder content (mentioned in the "Interactions" section) and increases not only interparticulate adhesion but also plastic deformation and stress storage in the matrix.

4.3.11 Crow's-Foot Cracking

Crow's-foot cracking is aptly named. Why a crow was chosen over a bluebird or a cardinal is unknown, but this type of crack resembles a bird's footprint. Generally these cracks have a defect nucleus with three cracks propagating from the defect as shown in the Figure 4.27. When a large population of crow's-foot cracks exists, it can be difficult to identify the defect correctly, since the cracks radiating from each nucleus run into each other. It also sometimes happens that a crack from one nucleus propagates into another nucleus (also shown in Figure 4.27). The vast majority of crow's-foot cracking need not happen.

It has been established that tape-cast layers, upon drying, experience lateral (x–y) shrinkage stress. In many tapes, this lateral stress is not enough to rupture the polymer matrix but results instead in slight curling of the cast layer. If, however, a starting point is given for a potential crack, the same tape could crack to relieve the lateral stress. In our experience, the vast majority of crow's-foot cracks have a defect at the nucleus. Defects causing crow's-foot cracking can be defined as "anything that is nonhomogeneous with the rest of the tape-cast layer." Analyses of crow's-foot nuclei have turned up bubbles, powder agglomerates, binder lumps, dust particles, skin lumps that let go from the doctor blade, and in one case a gnat. Filtration of the slip through an appropriate filter opening will remove most potential nuclei. As discussed in Section 4.1, filtering may or may not be appropriate for research trials but should be considered mandatory in production.

Filtration is not the only measure that should be used to prevent crow's-foot cracking. Since, in the real world, there is no such thing as a defect-free cast, the tape formulation should be designed so as to tolerate minor defects such as agglomerates or tiny interior bubbles. This "forgiveness" can be added to the tape by a few different methods, all of which target the ability of the matrix to dissipate crack-tip energy. Adding more binder to a normally acceptable tape formulation will provide a higher concentration of free binder, which will act to prohibit crack propagation. In the same manner, the addition of a Type II plasticizer will help to dissipate the crack-tip stress to prohibit crack propagation. Basically, the tape being cast should be able to tolerate point defects, even though efforts are being made to eliminate those point defects.

Some very thick tapes will display crow's-foot cracks, even though no point defect nucleus is present. Tapes with dry thicknesses above 510 μm (0.020 in.), even up to 3.8 mm (0.150 in.), take a much longer time to dry due to the increased diffusion distance from bottom to top of the cast layer. As mentioned previously, the polymer matrix tends to shrink more as drying slows. These very thick tapes are also normally cast from a high-viscosity and high-pseudoplasticity slip to avoid lateral spreading after casting. The combination of increased shrinkage and decreased particle mobility in the matrix (high viscosity) often causes increased lateral tensile stress in the matrix, resulting in crow's-foot cracking.

Thick tapes (> 500 μm) are also often cast with a lower binder content than thinner tapes. The lower binder content allows a greater extent of cracking because of the lower matrix strength. The cross-sectional area of these tapes is large (width × thickness); larger cross-sectional area results in higher tape strength, allowing lower binder content and thus lower cost. Since the binder can be decreased with these thicker tapes, and since increased binder content tends to decrease drying rate in an already slow-drying thick tape, binder content is reduced to increase drying rate. A third reason for decreased binder content in these thicker tapes stems from the burnout step prior to sintering. A thick, high-density sheet can require a significant amount of time for burning out the organic components. Any binder used in the casting process will have to be removed in further processing through a thick layer of densely packed material. Lower binder content in the tape-cast layer will shorten burnout times and lower firing defects by essentially reducing the amount of organic components to be burned off. The advantages of lower binder content must be balanced against the appearance of crow's-foot cracks in the tape. In both thin and thick tapes, the most common solutions to crow's-foot cracking are reducing solvent content (thereby reducing volumetric shrinkage), increasing binder content (thereby increasing matrix strength), and/or adding a Type II plasticizer to allow plastic deformation (thereby dissipating lateral tensile stress by bending the matrix instead of breaking it).

4.3.12 Lateral Sheeting

Curling and cracking are stress-relief mechanisms that exist in response to lateral shrinkage stress in the plane of the cast (*x–y* directions). The stress-relief mechanisms discussed so far have addressed the relief mechanisms in the *y*-direction or "cross-cast" direction, since it is often narrow and unbound. The narrow, unbound cross cast direction allows greater response to lateral stress. The "cast" direction (*x*-direction) also sees shrinkage stress but is typically under tension (carrier motion) and is a much longer distance (length of cast). The different forces on the cast in the cast direction tend to reorient shrinkage stress effects into the *y*-direction where relief mechanisms are readily available.

This, however, is not always the case. Tapes exhibiting a high percent age of lateral shrinkage, usually very thick tapes, will try to shrink in the cast direction as well as the cross-cast direction. Since the carrier film in the cast direction is restrained by the carrier drive system, and the length of the cast typically prohibits any edge curl, a tape with high lateral shrinkage may crack across the cast and shrink in the cast direction. This cross-cast cracking and separation is called "sheeting," since it results in short "sheets" of tape instead of a long cast. Sheeting can produce gaps between the sheets in excess of two inches (50 mm). As would be logical, this lateral sheeting behavior cannot take place without at least partial carrier release, since the carrier cannot conform in the cast direction to the lateral dimensional changes of the tape. Lateral sheeting is a rare defect to see, and it usually occurs only in very thick tapes (over 1 mm dry thickness) made with very fine particles (< 1.5 μm).

Sheeting cannot always be solved without major property changes in the green tape. Tapes that we have seen to exhibit sheeting behavior are thick enough to prohibit continuous casting and are limited to batch casting only. The thickness of the tape, and the resulting long drying time, require that the tapes be cast into the drying chamber and left for a significant period of time. Since continuous casting is not an option for these tapes, the sheeting of the tape can be easily tolerated without causing any major manufacturing problems. The self-release from the carrier surface permits lateral shrinkage, and the resulting densification of the cast layer often yields very high green bulk density and high GOOD density. The high density typically associated with tapes that display sheeting is sometimes more desirable than a continuous tape. GOOD is an acronym standing for "green oxide only density," which is calculated by dividing the mass of the oxides in the tape by the bulk volume. A more complete definition can be found in the glossary.

If sheeting is intolerable, it can be solved by increasing binder content as well as Type I plasticizer content. These tapes, being very thick, often have relatively low binder content due to their high thickness. Additional "free binder" increases interparticulate adhesion and also separates the particles, thus providing room in the matrix for plastic deformation. In order to dissipate lateral shrinkage stress by plastic deformation of the matrix, the binder must be made extremely flexible by increasing the Type I plasticizer / binder ratio, sometimes to 1.5:1 or higher. The addition of a Type II plasticizer can also help to avoid sheeting of a thick tape. Any of the mechanisms mentioned to avoid lateral sheeting work to separate particles and allow motion in the matrix. The result of particle separation in the green tape is a lower green density. The decrease in drying shrinkage to avoid lateral sheeting will eventually result in higher firing shrinkage during sintering and densification.

4.3.13 Rails

We have only seen rails in tapes exceeding 635 μm (0.025 in.) dry thickness or in tapes that displayed an extremely high tendency to skin. The formation of rails is the first drying defect that stems entirely from the dual-mechanism drying process explained in the beginning of this section (evaporation from top, diffusion from bottom). All of the previous drying behaviors can occur in a homogeneously drying tape layer, and all can occur with a near-uniform solvent concentration throughout the cast layer, even though this uniformity cannot exist. The formation of rails and the following defects can only happen when the top surface and the bottom surface of the cast layer dry at significantly different rates. It is important not to confuse rails with carrier dewetting or slip flow around the doctor blade. Minor edge dewetting from the carrier will cause thick beads along the tape edges, as will unwanted slip flow around the edges of the doctor blade (see Figure 4.20). Rails are found not at the outermost edge of the tape, but at least one-half inch inside the tape edges. The outermost edges of a tape displaying rails are generally tapered to zero thickness.

A tape exhibiting rails will have a somewhat uniform center thickness across much of its width, but regions of high thickness toward the edges of the tape. The extreme edges of the tape taper to zero as normal. The thick regions at the edges of the tape are usually less than twice the average center thickness but have been seen in some cases to exceed three times the center thickness.

The thick regions at the tape edges are not always readily apparent in the "as-cast" condition, since the thicker regions stem from the carrier film not lying flat on the bed. While the top surface of the cast layer remains relatively level and flat (except for edge curling), the bottom surface that is defined by the carrier surface is no longer flat. Since this drying effect is due to carrier curling, it is never seen when casting on carriers that cannot curl. Rails will never be seen when casting on glass, steel belt, or polymer riding on a convex bed. This drying phenomenon is only seen when casting on a thin, unrestrained, flexible carrier film over a flat bed. As solvent evaporates from the top surface of the cast, the gel structure is formed on the top layer. The top layer of the tape then begins to progress through the drying tree (Figure 4.26). The bottom layer of the tape, however, has not yet formed the gel structure and remains at Level 1 on the drying tree.

As the top layer continues to dry, edge curl lifts the carrier edges. With the carrier edges no longer flat, the bottom layer of the tape undergoes liquid motion under the force of gravity, pooling at the "corners" of the curled carrier edges. Upon the formation of the gel structure, the bottom layers of the cast are locked into position, creating a tape that is much thicker where the edges lifted than in the center.

In some very thick tapes, the top layer causes gull wing curling. The most extreme case of this rail formation occurs when the carrier forms the gull wing shape while the bottom of the cast is still fluid. In this case, the center of the cast can be very thin, while the thickness at one-quarter and three-quarters across the width of the tape is very high.

Rails can be classified as a drying defect, but it is not usually possible to solve the problem by changing the drying conditions. The solution to rail formation lies in the slip formulation. The dual-layer drying phenomenon is the root of this drying behavior, and the solution is to avoid, as much as possible, the formation of two distinct drying layers. In addition, the lateral shrinkage stress can be minimized and stress-relief mechanisms guided away from edge curling. Keeping in mind that the downstream process is affected by any changes or additions to the slip, the drying behavior and green tape properties must be balanced to create a tape-cast layer suitable for further processing.

The first drying stage of rail formation is the relatively fast gel formation of the top surface as compared to the bottom layers. A slow-drying solvent may be added to the system either in addition to, or in place of faster-drying solvents in the system. Other factors must be addressed, of course, such as the solubility of components, evaporation of components, and/or reaction with other components. Mentioned in the Solvents section of Chapter 2, cyclohexanone has been used for years in the industry as a skin retarder. We have also found xylenes (mixed) to work well as a slow-drying solvent. These types of solvents retard skin formation because of their low evaporation rates. As the top layer dries, a remnant of the slow-drying solvent on the tape surface allows liquid-phase diffusion of solvent from the tape bottom to the top evaporation surface, which allows the bottom to dry more quickly. Ideally, the binder should be soluble in the slow-drying solvent. This binder solubility is not mandatory, and nonsoluble skin retarders can work quite well, but binder solubility can help avoid residual porosity in the dried tape. A nonsoluble, slow-drying solvent interrupts polymer chain compaction while the polymer dries, thus holding the door (matrix) open so the bottom layers of solvent can get through. If the polymer is fully out of solution while a large amount of polymer-inert solvent is still present, the inert solvent will leave behind significant open volume (porosity).

The delayed gel structure formation on the top of the tape allows more solvent to diffuse to the top of the tape prior to any dimensional changes. The closer the bottom layer of the tape gets to gel structure formation before the edges curl, the less the extent of subsurface fluid motion causing rails. Switching solvents in a system from MEK-ethanol to xylenes-ethanol may avoid rail formation altogether. This is a case of slowing down the drying rate to increase the overall drying rate. The top-surface drying rate due to evaporation must be decreased in order to promote a faster liquid-phase diffusion, thereby allowing the bottom of the tape to dry more quickly. The top surface gel structure, or skin, can also be delayed by saturating the local atmosphere around the cast layer with vapor(s) of the solvent(s) used. This was mentioned in the introduction to this section and strongly discouraged due to safety concerns.

A second way to avoid rails during drying is to modify the stress-relief mechanisms active during the drying process. As in many of the cases discussed earlier, adding a Type II plasticizer will bolster plastic deformation under stress and reduce edge curling. This may or may not be adequate to prevent rail formation. In larger amounts, many tapes remain nearly or fully flat through the drying process, bypassing Level 4 stress-relief mechanisms altogether.

Another way to avoid rail formation is to hold the carrier film flat manually or mechanically. Many different mechanisms have been devised to hold carrier edges flat, ranging from the simple to the ridiculous. We both agree that carrier "edge hold-down" mechanisms are mechanical solutions to a chemical problem.

The last method for solving this type of drying defect is also the least effective. Some solvent-binder systems will generate tapes that lie flatter during drying as the temperature of the matrix increases. This stems from the glassy nature of the polymer. Raising the air temperature significantly above the polymer T_g may enable plastic deformation of the polymer matrix under the force of gravity, flattening the tape and carrier and avoiding the curling problem altogether. Unfortunately, the temperature increase tends to exaggerate other drying problems, including the solvent concentration gradient from top to bottom in the cast. Heating the casting surface can work toward keeping the tape flat, but it is only effective when the cast layer is in contact with the heated surface. As soon as curling begins, the cast layer is no longer affected by the "anti-curl mechanism."

A subsection of the rail defect during drying has been labeled "rail cracking." Some tapes will display cracking along both edges, approximately 25 to 50 mm (1 to 2 in.) in from the tape edges. Of those tapes that were analyzed for thickness profile, all displayed a greater thickness at the crack edges than across the body of the tape. The formation of these cracks is a combination of the rail defect and the center cracking described previously.

Stress-relief mechanisms increase in number and intensity as tape thickness increases. In tapes that display rails, it is common to see that the center of the tape is smooth and uncracked, while the thicker regions toward the edge display cracking. The center cracking defect will manifest along the centers of the rails, resulting in a wide, uncracked tape with thick, fractured edges and two very narrow, highly tapered strips along the edges of the cast. Whenever cracks are evident approximately one to two inches from each edge of the tape, with no other major defects, the tape thickness profile across the cracked region should be measured to determine if rail formation was the root cause of the cracking. Rail cracking is solved by avoiding rail formation altogether.

4.3.14 Wrinkling

Wrinkling is a very common occurrence in tape casting. While normally not seen in tapes under 50 μm (0.002 in.), wrinkling is usually one of the first drying defects encountered in a tape manufacturing process, providing the cast layer is thicker. Wrinkling, like rails, exists due to the existence of two-layered drying. A tape displaying wrinkling has a flat, smooth bottom surface (unless other defects occurred) but a nonuniform, wrinkled drying surface.

Wrinkling occurs in a cast layer when evaporation of the solvent from the top surface progresses more quickly than solvent diffusion, creating a very thin "dry" layer across the tape surface. The quick formation of this dry layer over a fluid support defines the size of the skin as equal to the tape surface area. Solvent from the underlying slurry then diffuses into the skin at a rate too slow to redissolve the skin. As solvent migrates toward the top surface, the polymer matrix of the skin layer swells, increasing in length and width. Since the energy needed to lift a small area upward is less than that needed to move a large area of skin, the skin layer wrinkles to provide additional surface area and accommodate matrix swelling.

Wrinkling can be a localized effect, created by a local zone of high-velocity air, or a systemwide effect if the overall air velocity is too high. Wrinkling is most often found in close proximity to air inlets and exhausts where cross-sectional flow area is smaller than in the rest of the system. The reduced flow area results in a greater linear air velocity, which can dry the tape surface much more quickly. Increased air speed over a local area of the tape surface not only replenishes a low-vapor-concentration atmosphere at the evaporation interface, but also reduces the local air pressure at the interface, further stimulating evaporation. One of us has seen concentric ring wrinkles around a 2-inch I.D. exhaust port that was improperly located. Wrinkling is quite common in tape casting processes, since all casting processes are designed for maximum production in minimum time. Since the root cause of wrinkling is drying of the top layer much too quickly, the increase of airflow to increase throughput can easily lead to wrinkling.

The solution to wrinkling is quite simple: slow the drying at the slurry/air interface. Whether accomplished by lowering airflow, lowering temperature, or avoiding high-velocity constrictions of the airflow path, the reduced evaporation rate will avoid the solvent-depleted layer and the wrinkling problem. Adding a "skin retarder" or even using a high-boiling-temperature major solvent component may not be very effective for preventing wrinkles. Any skin formed very rapidly is a source for wrinkles. In a xylenes/ethanol system (50/50 mixture), a polyvinyl butyral binder is soluble only in the ethanol component and will skin very quickly under high airflow, regardless of the slow-drying solvent. Even in systems of dual solubility such as ethanol/toluene, the very rapid evaporation of ethanol from a polyvinyl butyral binder can cause local skin formation even though the PVB is soluble in the slower-drying toluene.

Wrinkling is more prevalent in higher-binder-content systems due to the lower amount of solvent in the polymer solution. Lower binder contents have more solvent per unit of polymer and require a greater degree of evaporation to form a coherent film. Lower-binder-content films can tolerate higher airflows and higher temperatures before the onset of wrinkling. Some extremely high-binder-content systems with high-volatility solvents (MEK, acetone) display wrinkling of the cast surface even in room temperature, stagnant air. It is conceivable that wrinkling can also be caused by an airflow that is so high that it actually moves the cast layer of fluid like ripples. While this would certainly cause ripples in a cast film, we have never seen an airflow set high enough to work in this fashion.

One feature of wrinkling sets this drying defect apart from all of the other defects mentioned. Wrinkling of the cast surface is often transitory. A cast that displays mild to moderate wrinkling during the initial stages of drying may be flat, smooth, and uniform when fully dry. Since the mechanism responsible for this defect is the fast formation of a thin skin layer above the fluid cast layer and the resulting compressive force from swelling, continued diffusion of the solvents through the thin skin layer can relieve the stress and rearrange the polymer matrix (or break it down fully), resulting in a flat layer. Tape casting is generally a continuous process in which the cast layer will only see a localized high-flow area for a short period of time before moving away from it. Any wrinkling caused by the high-flow area will probably have a chance to "heal" before the entire polymer matrix is permanently established. Not all wrinkling "heals," but it is a common occurrence.

A side effect of wrinkling, and perhaps other two-layer drying defects, is a significant gradient of binder from top to bottom in the tape. The gradient has been seen to be greatest, in fact, through the first few mils (10 to 70 μm) of the drying surface. Tapes that display signs of two-layered drying (solvent-depleted zone, skinning) most often have a thin, binder-rich layer at the drying surface. The initial skinning of the cast layer creates a high-binder-concentration layer across the tape that lasts through the entire drying process. Diffusion of solvent from bottom to top of the cast layer also carries binder to the top surface. For this reason, these tapes tend to have different bend radii for top and bottom surfaces. Bend radius is described later under "Tape Characterization." The high-binder-content top layer logically displays higher tensile strength than the binder-poor bottom layer.

4.3.15 Mud Flat Cracking

As with orange peel, "mud flat" or "mud-flat cracking" is named for the resulting tape surface appearance. Tapes showing mud-flat defects resemble parched soil or dry lake beds, with random surface cracks creating small islands of material. A tape-cast layer can exhibit the same type of cracking. The difference between mud-flat cracking and extensive crow's-foot cracking as previously described is the depth of the cracking. Mud-flat cracking differs in appearance from center cracking and crow's-foot cracking in that the cracks do not go through the entire thickness of the tape. Mud-flat cracks go through the top layer(s) of the cast layer but stop short of the carrier surface, leaving a fairly continuous tape.

Mud-flat cracking is a product of two-layer drying. As mentioned previously, the top layer(s) of a cast film dry to the gel structure more quickly than the lower layer(s). With this drying behavior, combined with the glass-like matrix formation behavior of the polymer, tensile stresses form throughout the drying surface, causing the mud flat appearance. The polymer matrix has been compared to a glass matrix many times throughout this book due to the numerous similarities between the two.

As a glass cools through T_g, the glass smoothly transitions from liquid to solid. Cooling at a faster rate will generate a larger, less dense glass, while slow cooling yields a smaller, denser glass. This is the foundation of glass tempering. In a ceramic tape, the cast layer is not cooled through T_g but, conversely, effective T_g is raised through environmental temperature. The structure formation follows a relationship very similar to glass cooling. Upon initial matrix formation, a top "solvent-depleted" layer transitions from liquid into a solid matrix, establishing dimensions in length and width. As solvent diffuses through this layer, however, the presence of T_g-lowering solvent allows continued motion of the polymer matrix, leading to higher density and lower volume of the surface layer. This continued shrinkage and densification after defining a volume creates tensile stress in the top layer.

The bottom layer(s) of the cast dry at a slower rate than the evaporation layer. The end result of this slower drying (similar to slower cooling of a glass) is a higher density and correspondingly lower volume per unit of slip. This higher shrinkage of the bottom layers combats the tensile stress of the top layer, essentially pulling the upper matrix together. This mechanism, when the cast is on a fixed carrier, can lead to the curling and cracking mentioned earlier. If, however, the tensile stress created in the top layer reaches a critical point prior to the compression imposed by the lower layers, the top surface of the tape will literally rip itself apart. After the top layer has separated to relieve stress, the lower layers of the tape hold the newly formed "islands" of material together, maintaining the cast in a continuous film. The end result of this course of events is a continuous, or nearly continuous, cast with frequent, random cracks that progress only partway through the thickness of the cast. In some cases, the only visible remnants of mud-flat cracking are top surface discolorations and slight thickness variations around tear lines. In other cases, measurable separations are left in the top surface of the dry tape.

Mud-flat cracking can almost always be solved, or at least lessened, by the addition of a Type II plasticizer. Plastic deformation provides a tensile stress relief mechanism for the top surface in place of tearing of the skin. The addition of a skin retarder can also alleviate mud-flat cracking by preventing or delaying the formation of the surface skin. Delaying skin formation is equivalent to slowing the drying rate of the top layer (slowing the cooling rate for a glass) and allowing the top surface to form a denser, lower-volume layer, thereby lowering the tensile stress encountered later in the drying process.

Mud-flat cracking is more prevalent in low airflow drying conditions, so lowering airflow is not generally effective for preventing this defect. The mud-flat result is seen more often in low-airflow conditions because higher-airflow conditions will tend to crack the tape through the entire thickness. It is substantially correct (though there are exceptions) that mud-flat cracking is a very limited onset of cracking, in general, both along and across the body of the cast. This being the case, increasing the binder content is also a viable solution to mud-flat cracking, as it was for center cracking.

4.3.16 Orange Peel

This drying defect is named appropriately. This defect causes the top surface of the tape to resemble the skin of an orange. Tapes exhibiting orange peel surface structure have a high degree of porosity at the drying surface, presumably due to pore channels left behind by the evaporating solvent. Presumably caused by mechanisms similar to those that cause wrinkling and mud-flat cracking, orange peel leaves small craters in the tape surface that can remain on the surface of the fired part.

Orange peel is solved in much the same way as the previous two-layer drying defects. The top surface drying of the cast must be slowed in order to allow ample liquid diffusion from the body of the tape to the evaporating surface. Orange peel as a drying defect is not limited to the tape casting industry, but is also found in other coating industries. Orange peel defects seen in shellac coatings are evidence that the drying environment is too hot. The high heat speeds the drying of the exposed layer, leaving pore channels behind as the underlying vehicle migrates to the drying surface.

4.3.17 Warping

Warping or lateral curling (curving) of a tape-cast sheet is not seen regularly. The term warping as applied to a green ceramic tape does not describe vertical distortion of the tape and is not synonymous with curling. The cause of warping is differential shrinkage of various areas of the cast layer. Warping can only stem from the existence of severe nonuniformity in either airflow, temperature, or slip. Since airflow is typically required by safety regulations to fall well into the "turbulent flow" region, nonuniform airflow is an unlikely cause of lateral warping. Slip inhomogeneity due to settling, poor mixing, incompatibility of chemical components, or poorly dissolved binders is typically to blame for warping.

Differences in binder content along two edges of a cast layer will cause distinctly different lateral and vertical shrinkage. When two sides of a cast layer shrink by different amounts, a stress gradient will be formed from side to side in the tape. A very large tension difference is required for the net force to break the tape–carrier bond and physically move the drying tape sideways. While rarely seen, warping may occasionally be observed with polymers that tangle in solution, forming highly localized "gel pockets." When these high-binder-concentration "gels" are metered by the doctor blade, they form large areas of high binder concentration as compared to the surrounding binder-poor regions, which results in warping upon drying.

4.3.18 Other Interactions

Some other drying phenomena are unique enough to fall into the category of "Other." Three of these are addressed in this section. There are certainly other unique drying situations that have been developed or discovered for various applications that are not included here. The three occurrences mentioned here include swelling of the polymer, beneficial carrier/slip interaction, and one other that is not yet fully understood.

One of the systems used by the authors, and described in sample recipes in Appendix 1, is the xylenes/ethanol dual-solvent system with polyvinyl butyral as the binder. PVB is soluble in ethanol, and xylenes is an effective skin retarder. PVB, however, is not fully soluble in xylenes. Trade literature on PVB describes the xylenes/PVB interaction as a swelling of the polymer.[31] Since xylenes is a much slower-drying solvent than ethanol, a stage exists during the drying process when only xylenes remains in the polymer matrix. Since xylenes "swells" the polymer, the matrix will shrink upon evaporation of the xylenes, effectively increasing green density. This "self-compaction" of a cast tape during drying has not been quantified but may be an interesting area of study in the future.

The Western Electric process for producing supersmooth alumina substrates included the use of cellulose triacetate as a carrier film. It was found that cellulose acetate would react to some degree with the solvents being used (ethanol/trichloroethylene). The carrier film would swell in contact with the solvents and then shrink along with the cast, helping to relieve lateral shrinkage stresses and stress gradients in the z-direction. It should be mentioned that, even with this novel carrier interaction, the casts typically exhibited gull wing curling, anyway. This gull wing curling was tolerated in the production process.

Perhaps the most interesting interaction noticed is interesting because it goes against common sense. Many ceramic formation processes limit the amount of organic additives in an attempt to increase green density. This is also true of tape casting. Binder content is often kept at low concentrations in order to achieve higher ceramic particle loadings and resulting higher densities. Mentioned briefly in the Section 2.4, an increase in acrylic binder concentration can sometimes significantly increase density. We have observed marked increases in both green bulk density and GOOD density with increases in binder content. Binder concentrations as high as 20 wt% in alumina systems have generated green bulk densities well in excess of 60% of theoretical. Originally thought to increase density by filling pores with excess polymer, this densifying action of excess polymer also increases the oxide-only density, which only accounts for the ceramic material in the matrix. Higher GOOD densities show that the ceramic particles are actually closer together. The mechanisms that cause this density increase are not fully understood at this time (by the authors).

CHAPTER 5
Further Tape Processing

Once the tape-cast product has been fabricated in either sheet or roll form, it is ready for evaluation and characterization before being processed into a useful shape or part. This chapter will cover the green tape characterization techniques and procedures that are used in industry, as well as procedures we have instituted in our laboratory. The remainder of the chapter will address the downstream processing steps followed in the final shaping procedures to produce parts for specific products. These procedures include: blanking, punching, calendering, and lamination. First we will review the tape characterization procedures and techniques.

5.1 TAPE CHARACTERIZATION AND ANALYSIS

After the cast layer is dried and removed from the machine, most tape producers take advantage of the natural pause in the manufacturing process to inspect their work. Standard quality control statistics and statistical process control data can be generated from measurements of green tape properties. The measurements and data-gathering techniques for green tape are generally bulk and visual measurements. Some manufacturers will include such tape performance tests as strength, yield to failure, and smoothness tests at this stage. These "green performance" techniques are not, however, an integral part of every manufacturing process. It should be obvious why. The green tape is merely an arrangement of powder particles configured to yield the desired fired part. The binder, plasticizers, dispersants, and other organic additives are about to be vaporized in the kiln. Measuring the strength of the green tape yields data on a polymer matrix that is about to be destroyed. This is not at all to say that all quality control methods are useless, simply that quality control should be centered on properties that will help yield a higher-quality part after firing.

Some techniques commonly used for tape analysis are described in this section, along with some less-common test methods. We have developed some tests for tape property analysis that are also described here.[1] Our tape property test methods were developed outside a manufacturing mindset. We developed them to gain insight into the behavior of the polymer matrix, the effect of different additives, and the relative utility of various organic systems. Green tape properties such as tensile strength, bend radius, and stiffness may be of concern in a manufacturing environment as well. Further processing of the tape (downstream processing) may include automated tape handling, roll-to-roll printing, blanking, or other similar handling. These processing steps require properties such as strength, flexibility, and/or stiffness to remain within certain tolerable ranges. In these circumstances, verification of green tape properties would be valuable. Whatever the reasons tape characterization is performed, the following tests are included in the analytical arsenal.

5.1.1 Defect Analysis

Taking advantage of the characteristically thin *z*-dimension and continuous polymeric phase in a green tape, perhaps the most common defect analysis technique is light transmission analysis (LTA). This technique compares the transmission of light through the tape in various areas of the tape and helps to identify bubbles, streaks, binder-rich areas, debris or dust, wrinkles, changes in thickness, and agglomerates. Since the ideal tape would be homogeneous in chemical composition and in thickness, the ideal tape would transmit light equally through all areas of the tape. Any deviation in light intensity using this procedure can be marked as a defect.

Many types of equipment are used for light transmission analysis, all of which center around the same common principle. A source on one side of the tape transmits light through the tape, which is measured by the detector of choice on the other side. Quite often, the detector of choice is the human eye. Based upon what is seen, defects can be marked and either removed from the tape or simply avoided during further processing of the tape. Light transmission stations can be part of the casting process (in-line) or a separate operation (off-line) between casting and the next manufacturing operation.

Light boxes are available, typically from art supply distributors, which provide a large area of uniform light. These light boxes can be installed as part of the casting surface between the end of the covered drying chamber and the product take-up-reel. Depending upon the speed of casting, this arrangement can be used to identify and mark defects during the casting operation. When casting speeds are too high for in-line analysis, or too slow for in-line LTA to be efficient, a separate station can be used for roll-to-roll analysis of the cast. This discrete quality control station, while taking more time than an in-line process, allows the inspector to adjust the inspection rate or even halt tape motion for analysis or marking. An in-line inspection system is not variable in feed rate since, from the physics section, carrier speed is one of the four primary variables controlling tape thickness.

Light transmission analysis is used not only in roll-to-roll applications, but also as a quick check as processing continues. Light boxes can often be found just before the printing station to check for surface debris, before lamination to check for debris, after screen printing to observe pinholes in the printing, and so on. Tape thicknesses over approximately 200 μm (0.008") will generally not transmit enough light for LTA to be a comprehensive quality control method. Transmission can still be used though, even with thick tapes up to 3800 μm (0.150"), for locating bubbles or pinholes. Each manufacturing process, and sometimes each production line, must determine its own acceptable level of defects. A pinhole 3 μm in diameter may not significantly affect a functionally graded composite or metal matrix composite, but would cause total failure of a multilayer capacitor.

Analysis of defects in thicker tapes (> 200 to 250 μm [> 0.008 to 0.010 in.]) tends to be limited to surface defect identification. Since light cannot transmit adequately through the thicker tape, light is instead reflected against the tape surface to examine it for inconsistencies. This method is very efficient for locating agglomerates, debris, cracks, and large pinholes. Analysis using this method is most useful when the drying surface of the tape is smooth enough to be reflective. Nonuniformities are located using this method by identifying "variations in the glare."

Another useful light reflection technique for use with thicker tapes is low-level shadowing. Light striking the tape from a low angle will topographically shadow the surface, creating strong differences in contrast according to surface texture. In a manner similar to topographic analysis using a backscattered SEM image, surface features can be identified and marked as defects. This low-angle lighting technique is a fast, efficient method for locating streaks, bumps (agglomerates or debris), pinholes, orange peel, bubbles, and mud-flat cracking. This method may, however, miss small crow's-foot cracks or other defects that do not create large topographic changes.

5.1.2 Thickness / Thickness Uniformity

The most important characteristic to be measured, the tape thickness, is measured and interpreted in a variety of ways. Various instruments are available for direct measurement, as well as for measurement of other thickness-related variables. Average thickness is usually tolerated within a range determined by the final product needs, while thickness uniformity is nearly always minimized with all due diligence. While it is a physical impossibility, a zero-variance tape is the goal of every tape manufacturer. (It should be mentioned here that some specialty casters are intentionally casting thickness gradients into tapes. These are certainly exceptions and not the rule.) Section 4.2 covered methods of making a uniform tape and the reasons behind them. In this section, we will examine methods for quantifying the results.

The most direct way to measure tape thickness is with a micrometer. Shown in Figure 5.1, the standard flat-faced micrometer is used to measure maximum tape thickness over the area covered by the micrometer face. Perhaps the most commonly used micrometer face in the tape casting field has a 1/8-inch radius. Micrometer faces come in various sizes and shapes, each with a specific usefulness, but the low yield stress commonly seen in green tapes precludes the use of radius-end or point micrometers, since they tend to indent the tape during measurement and thus yield an inaccurate reading. As the flat-face radius increases, pressure applied by the micrometer during measurement decreases, making the reading more trustworthy and repeatable. The measurement itself, however, would be the maximum thickness within a larger area and thus would be less useful. The 1/4-inch-diameter flat-faced micrometer, therefore, is simply a happy medium between accuracy, reliability, and simplicity. Figure 5.1 also shows that the flat-faced micrometer can be used incorrectly and give false data. We realize that this book is not a tutorial on the use of a micrometer, but we have seen too many parts scrapped,

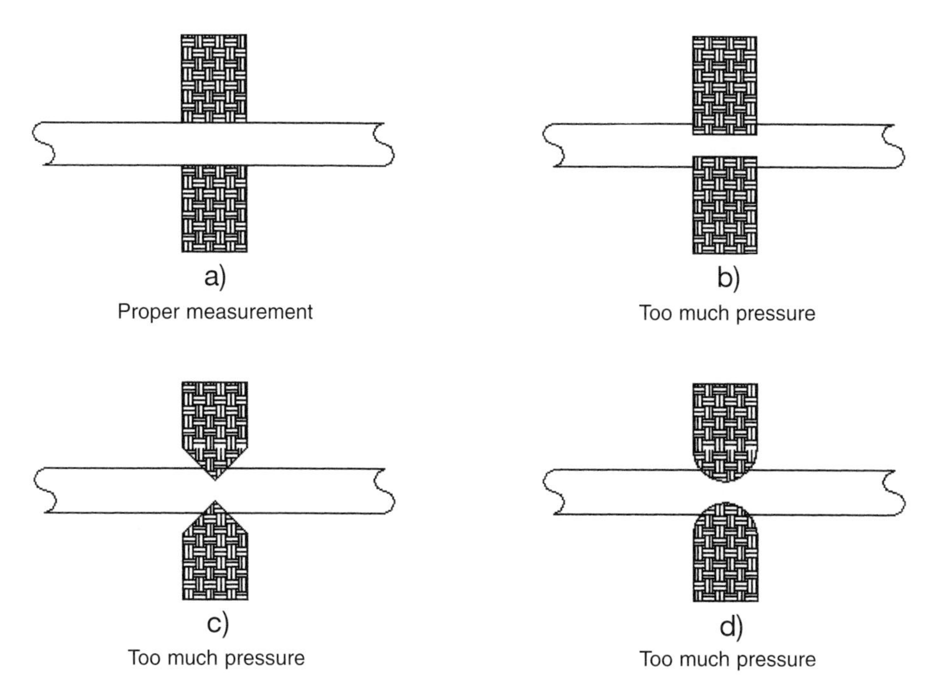

Fig. 5.1 The standard flat-faced micrometer is used to measure maximum tape thickness over the area covered by the micrometer face.

and too many projects questioned due to improper measurement to leave it unaddressed here. A somewhat safe rule of thumb is: If you put a dent in the tape, you probably pressed too hard.

Thickness measurement interpretation is also handled in a number of different of ways. While most manufacturers routinely use weighted mean averaging, others find it more efficient to use an "all points within" approach. Since the final product needs will dictate the needs of the green tape, a certain minimum and maximum green layer thickness can be established to yield a good fired part. With these parameters set, as long as all points measured on the green cast layer fall within these limits, the tape can be considered a good cast, regardless of the mean thickness or standard deviation.

Profiling of the cast tape, creating a numerical and/or visual model of tape thickness over the tape width and a substantial length, can also be a useful analysis technique. Long-term trends or property cycles can be identified and addressed by this type of characterization. Periodic filling of the reservoir head during continuous casting will be seen in the green tape thickness by this type of analysis. Using thickness profiling, we were able to attribute one cyclic thickness variation to the chain drive system on the conveyor. Switching to a belt drive removed the variation. Profiling can also give necessary information as to the effects of the slip feed system and locations of slip addition, carrier speed uniformity, defects on the casting blade, carrier feed, carrier tension uniformity, the degree of levelness of the casting bed, and a variety of other variables that directly affect the quality of the green cast layer.

Profiling in this manner is not limited to thickness alone. A good case can be made for the importance of mass uniformity in many tapes. Systems actually exist in which mass uniformity is a more critical parameter than thickness uniformity. Regardless of the system, mass uniformity within a cast layer has great importance, since it affects shrinkage uniformity, density gradients, warpage, electrical properties and the like.

Mass uniformity measurements can be done by a number of methods. Two methods used to measure mass uniformity are X-ray transmission and weighing. X-ray transmission mass measurement is a method similar to that used in wet thickness monitoring in that mass is measured by its effect on X rays (typically gamma). Where wet thickness monitoring often uses backscattering, however, this profiling technique measures transmission. Benefits of X-ray transmission analysis include a very small measurement area which allows the identification of thin streaks of low mass along the casting direction. One limitation of this technique is detector sensitivity. It may be necessary to stack multiple layers of tape in order to yield meaningful results. Low detector sensitivity limits this technique to medium- or long-range trend analysis.

Direct weighing of small samples is a more accurate and quantitative analysis method, but is also more time-consuming. Blanking samples from the tape, or carefully cutting identical pieces, and recording both location and weight can yield a good picture of mass uniformity throughout a cast tape. This is a destructive measurement technique and thus would only be performed periodically along a continuous cast. More general trend information can be nondestructively measured by blanking or cutting out larger areas of tape (called cards, bars, sheets, or blanks, depending on the industry) and weighing each before further processing. This quality control technique, commonly called sheet mass, is a quick and simple technique that is easily integrated into a production line. With thickness measurements made on each sheet as well, green bulk density can also be tracked.

Green bulk density (GBD), also called envelope density,[2] is a measure of the mass per unit volume of a material and gives insight into how well the components are packed. When a high-density fired product is desired, high GBD is also desired. Simply dividing the measured mass by the measured volume will yield the GBD of a particular sample. This information is vital to quality control of a cast, since it is directly related to the distance between ceramic (or other) particles in the tape. Changes in interparticle distance have a resulting effect upon firing shrinkage and can also affect fired bulk density (FBD) in many cases.

Green bulk density, while directly affected by interparticle distance, is not an accurate measure of particle packing. Variations in other parameters also affect GBD and can be misleading. Since a dry green sheet includes organics and porosity in addition to the cast powder, changes in these can cause dramatic fluctuation in GBD. A higher binder content and lower percentage of porosity will effectively increase GBD, even though the particle packing remains constant. GBD is an effective quality control parameter for a production line that consistently casts the same tape formulation. When changes in particle size, distribution, binder loading, or tape thickness are common, a slightly different approach may be beneficial.

Since the primary governing parameter for shrinkage and fired density is the proximity of the powder particles to each other, a method that relies solely on this parameter may be more effective. We have not found this technique described specifically in previous literature and thus have named it "green oxide-only density." The technique is simply a mathematical manipulation of GBD measurements to remove the contributions to density from any fugitive material (binder, dispersant, etc.). The green oxide-only density (GOOD) is a measure of the mass of powder per unit volume. This number can be obtained by calculating the ratio of powder weight to total non-volatile weight (including the powder) from the slip recipe. The GBD measurement can be multiplied by this ratio to yield the GOOD. Oxide-only (or powder-only) density is an effective technique for quantifying the effect of different binder concentrations, lot-to-lot variations of organics, and comparisons between different powders.

The remaining characterization techniques covered in this chapter are methods of characterizing the tape properties themselves and do not translate into fired part performance. As mentioned earlier in this chapter, techniques have been developed to test the properties of the green tape itself. Parameters such as strength, flexibility, and plasticity are meaningless as they pertain to properties of the fired part, but they may be critical for green handling, particularly with automated handling systems. The strength of the green tape is of great importance when a roll-to-roll handling system is used for slitting, blanking, printing, or via punching. Flexibility is an important issue when single layers are handled manually for blanking and for rolling. Plasticity is a very important issue during free handling, carrier removal, and lamination. Both cost and time can be saved by knowing that a strength problem exists before the tape breaks during blanking, or that a plasticity problem exists before the pieces show memory defects after firing.

5.1.3 Tensile Strength

Many tapes are further processed with automatic or semiautomatic handling equipment. The tensile forces applied to the cast tape, specifically when the tape is processed separate from the carrier film, require the tape to have a certain level of tensile strength to avoid damage. Specifically, the tape must have a yield strength greater than the maximum tensile force applied. The green tensile strength required is most often empirically derived, since the quantitative measurement of tensile forces applied throughout the handling process is difficult and time consuming. Further, most thicker tapes (> 200 μm [0.008 in.]) have much more than adequate strength due to their increased cross-sectional area. Standard tests exist, however, for measuring tensile strength that can be used for measuring green tape tensile strength.

Standard tensile strength tests, similar to those for leather (ASTM D2209-95) or for metals (ASTM E8M-98), are also applicable to green tape. In the field, this tensile testing method is commonly called Instron testing after a manufacturer of this type of testing equipment. The loads involved, however, should be expected to be much less than those for metals or leathers. Tensile testing of tape, typically performed on I-shaped samples, or "dog bones" as they are commonly called, will generate data not only on yield strength, but on ultimate strength, yield to failure, and compliance as well.[3,4] The limitation imposed by the green sample, however, is the generation of critical flaws introduced during sample preparation. Any edge nonuniformity existing in the cut or blanked sample can create a stress concentration, thereby lowering overall tensile strength. Great care in sample preparation can minimize this phenomenon, but the normally high strain and low strength of green ceramic tape as compared to other materials magnifies the effect of this type of critical flaw.

Another area of primary concern with this type of testing is centered on the sample holder. Alteration of the sample during mounting and by the sample holder during testing due to the normally high compliance of tape can result in sample failure at the sample holder, thereby invalidating the test. Standard vise-type grips or clamps are quite often unsuitable for unfired tape. Methods of avoiding this problem have included rubber-lined or sponge-lined grips, and pneumatic grips for pinching the sample ends.

The last major area of concern for this type of testing is sample alignment. The flexibility of a tape-cast sheet makes perfect alignment difficult. Sample misalignment in this type of test, especially with a highly compliant material, can dramatically affect the resulting data. Sample misalignment concentrates stress on one edge of the part and reduces the effective width of the tested piece.

5.1.4 Bend Radius

The ability of the tape to bend under an externally applied force without cracking is important in any process where bending of the tape is likely. Roll-to-roll processing, continuous casting onto a roll for storage, and manual handling of the tape all require the green tape to bend. Bending of any material requires differential strain on the two surfaces. Bending a layer elongates the outer (convex) layer, compresses the inner (concave) layer, or both. A "plane of zero strain" exists, usually within the layer, outside of which the layer is in tension and inside of which the layer is in compression. The densely packed structure of the tape will always fail first in tension. The ability of a tape to bend, therefore, is limited by the ability of one layer to elongate without elongation of the opposite layer.

Tensile testing methods will generate the quantitative elongation data desired, but can be time-consuming and expensive. Figure 5.2 shows a simple test for quantifying the elongation of a cast tape.

This test (mandrel bend test) must be performed with care to maintain the "plane of zero strain" at the mandrel/tape interface. This can be accomplished, or closely approximated, by rolling the tape onto the mandrel from a flat surface. Assuming the plane of zero strain at the mandrel interface, the percent elongation can be calculated using the equation shown in Figure 5.2. Close analysis of the outer surface is then made for cracks. This defect inspection should be done while the tape is still on the mandrel. This type of testing yields a range within which the maximum elongation will fall. The maximum elongation before cracking will differ for the top and bottom of most tapes. Elongation will also be a function of temperature due to the thermoplastic nature of the binder polymer. This test is most useful and applicable to tapes thicker than 125 μm (0.005 in.).

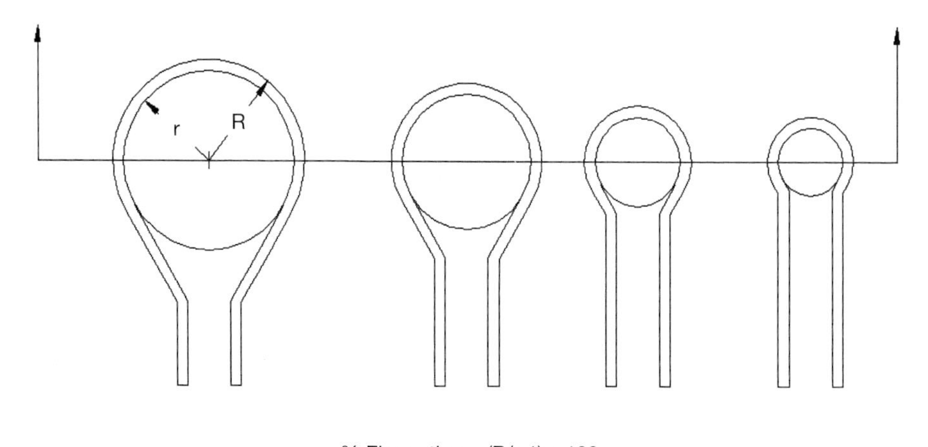

% Elongation = (R/r-1) x 100

Where R = r + t

and t = thickness of tape

Fig. 5.2 A simple test for quantifying the elongation of a cast tape.

5.1.5 Deflection

Part of making a product by tape casting is handling the dry green tape. This handling, especially if it is done by hand, can be more convenient or less convenient depending on the stiffness of the green tape. Single-sheet pickup and handling is more convenient with a fairly stiff tape. Stiffness, or in this case Young's modulus, can be calculated by the tensile test mentioned previously. A fast and simple test has also been devised to quantify relative deflection from cast to cast.[1] This test is designed to measure the extent to which a tape will deflect (droop) under its own weight without an externally applied force.

Shown in Figure 5.3, this test measures the extent to which the ends of a known length of tape will droop in a given period of time. In this test, a rectangular sample of tape with a length:width ratio of at least 10:1 is centered lengthwise on a flat support. The support plate is notched in the center. The plate is then lowered over a small-radius cantilevered dowel, ensuring contact with the center of the tape sample. Vertical deflection of the tape is then measured visually by sighting along both ends of the tape and comparing to a vertically mounted scale. Results from the "droop test" can be expected to vary with testing temperature, sheet mass, tape thickness, tape drying conditions, and time on the mandrel.

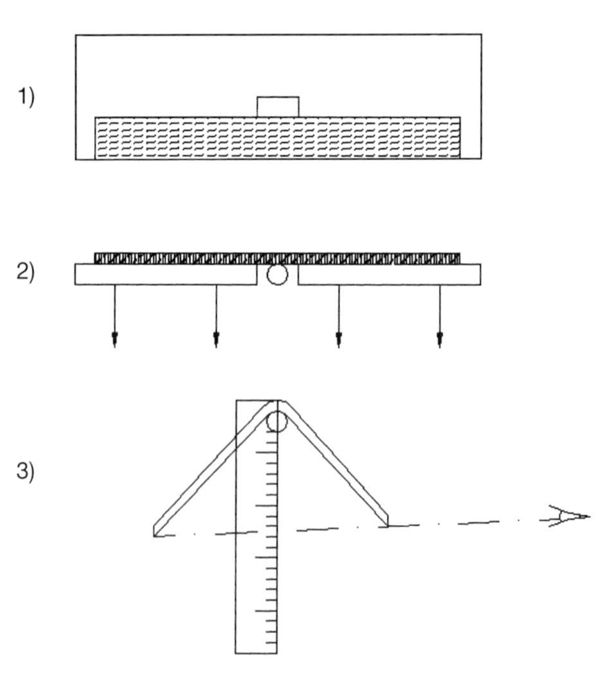

Fig. 5.3 A test measuring the extent to which the ends of a known length of tape will droop in a given period of time.

5.2 SHAPING

The next step in the processing of tape-cast materials is to punch the desired shape in the area defined by the x and y directions. Most production facilities use punch-and-die sets in either a hydraulic or a mechanical press. The shaping process usually is divided into two separate functions that are sometimes combined in simple parts. These procedures will be described in the next sections.

5.2.1 Blanking

This is the punching process that forms the outside shape of the desired part. Most blanking tools are made of either a hardened steel (usually A-2 or D-2 tool steel) or a carbide material, depending upon the volume of parts to be manufactured. Most tools using carbide only have the carbide on the cutting edges, not on the entire tool face. Large internal holes or other shapes can also be punched into the part as part of the blanking operation. In some cases, specially designed punches with recessed regions and carbide cutting edges are used so that contact with the tape is only made at the very perimeter of the shape being punched. This technique is used by the manufacturers of thin-film substrates when the surface quality of the as-fired substrate is critical to the final use of the product. For these substrates, the thin-film circuits are deposited directly on their surfaces and any defects such as scratches can disrupt the conductor or resistor patterns and cause circuit failures. Some sort of cleaning, whether it is a vacuum cleaner or a strong air jet, is essential to keep the punch and die clear of debris between punching strokes. For most ceramic substrate applications, the punched part is a simple square or rectangular shape, but a wide variety of other shapes can be generated using the blanking process.

The design of blanking tools is an art in itself. For most ceramic tape punching, the clearance between the punch and the die is on the order of 38.1 μm (0.0015 in.), and most corners have a small radius on the order of 0.254 mm (0.010 in.).

Tools are designed in two ways:

(1) as a "spanking" type, where the punched parts are pushed out through the bottom of the die into a collection tray

(2) as a "retraction" type, where the part is pushed back up under a stripper plate to a stopped position, where it is removed manually.

The "spanker"-type tooling is much more compatible with a continuous production process. An automated "spanker"-type blanking process was described by ALCOA in a recent patent.[5]

Other blanking procedures have been described in the literature. The American Lava Corporation[6] described a roller blanking procedure in which the longitudinal cuts (those in the casting direction) were made by cutting rollers and the transverse cuts (cross-cast direction) were made by a synchronized cutting blade. The American Lava process was done on the carrier, and the parts were stripped directly onto refractory setters just prior to sintering. IBM Corporation[7] mentioned a roller blanking procedure on stripped-tape product. Details were not provided, but the blanking was probably done as a single operation, since there was no carrier involved. Neither of these procedures involved the simultaneous introduction of via or through-holes. The IBM process had a separate step for via punching, but we believe that the introduction of alignment holes was done as part of the blanking process. Many of the blanking procedures do include the punching of a series of alignment holes around the periphery of the blanked part for use during subsequent processes such as via hole punching, screen printing, and lamination. The alignment holes are usually about 6 to 8 mm (about 0.25 in.) in diameter and are arranged so that two or three holes can be used for each operation (and only for that operation)—that is, a set of holes is used for the via hole punching, another set for the screen printing, and so on. If the manufacturing process requires the dicing of many small square or rectangular parts from the tape, an automated hot-knife blade cutting system is recommended.[8] The "green" parts produced by this technique usually are on the order of 3 to 4 mm (< 0.25 in.) in size. A piece of equipment has recently been described [9] that was developed specifically for MLCC (multilayered ceramic capacitor) production, where the tapes are less than 10 μm (0.00039 in.) thick. For tapes this thin, it is essential that all of the prelamination procedures be performed while the ceramic tape is still adhered to the carrier (Mylar or other). On this machine the screen-printed tapes are loaded as a roll, which is then continuously cut longitudinally (along the cast direction) on both sides. The tape then passes to the second station, where the transverse cuts are made and the carrier is stripped from the tape while it is held by a vacuum transfer plate. This transfer plate then moves to the third station, where the vacuum is released and the cut tape is stacked to form a 100 (or more)–layer structure. The last station on the machine performs the lamination process. The size of the blanked parts is approximately 150 × 150 mm (5.9 × 5.9 in.). These parts could not be handled in any manner other than by a completely automated system.

5.2.2 Hole or Via Generation

The next step in part fabrication is the generation of a hole pattern in the tape-cast material. These holes are used to interconnect layers in a multilayered ceramic package or to interconnect the two surfaces of a substrate. The holes are generally circular, but they can be most any other shape. This section will address the techniques and equipment used in the via generation process.

Hundreds of via holes can be punched simultaneously with the blanking operation or at a separate station from the blanking. Usually, repeating patterns of holes are punched in a group, although computer-controlled single punches are sometimes used. The advantage of using a single punch is that it is only necessary to change a single punch pin if it breaks. This becomes a less costly procedure than shutting down an entire blanking/punching tool for repairs in the event of a single pin breaking out of many. When punching multiple holes, the tool-cleaning procedures outlined previously become very critical, since the "slugs" of tape have to be removed between strokes, especially from the die cavity into which they are being pushed during the operation. A technique that we have used successfully is to use a partial vacuum in the chamber under the die to assist in this slug removal. An air jet that produces a Bernoulli effect is usually sufficient.

Automated equipment is now manufactured for production punching of micro-holes in tape-cast product. Most of these machines use a computer-controlled punch head with reported punching speeds of up to 10 holes per second on 2.54 mm (0.100 in.) movements. The reported hole sizes range from 0.102 mm to 9.525 mm (0.004 to 0.375 in.). Square holes, triangular holes, and oval-shaped holes are also within the capability of these machines. The reported positional resolution is 2.54 μm (0.0001 in.) and the positional accuracy is ± 5 μm (0.0002 in.).[10]

Via holes larger than > 0.5 mm (0.020 in.) are usually very easy to form, since materials such as tungsten carbide can be used as punching tools. For smaller holes, in the range of 0.25 mm or less, tungsten carbide may be too brittle, and other materials such as borided tool steel should be used. There are several companies that boride the tool steel punches that have been machined to size.[11,12] This process yields a punch with the surface hardness of a carbide but with some flexibility.

The precision location of holes in the final sintered part is very difficult to achieve, since the shrinkage in the x and y directions may be slightly different in normal tape-cast materials. In addition, the shrinkage in most tape-cast materials is usually quite large, around the order of 18 to 20 linear percent. An excellent paper that deals with this subject in detail was published by Piazza and Steele.[13]

Punch and die design is critical for tape-cast materials. The design criteria are different from those for standard metal punching tools, so they will be reviewed here:

- Use hard punches and a softer die material. Tungsten carbide punches and borided steel punches are used with a tool steel hardened to Rockwell 60.

- Design the punch so that it extends through the die into a larger chamber (or cavity) when the tool is closed. This will push the tape "slugs" completely through the die.

- The punch-to-die clearance should be 0.038 mm (0.0015 in.) on all sides. For example, a 0.305 mm (0.012 in.) diameter punch should be used in a 0.381 mm (0.015 in.) diameter die.

- Use a stripper plate to hold the tape down during punch withdrawal to prevent hole chipping.

Because of irregularities in most tapes, narrow punches tend to flex sideways far enough to strike the die during the punching stroke. Therefore, brittle materials, such as carbides, should always be mated with relatively soft materials such as hardened tool steel. This prevents premature wear and breaking of the brittle tools. Most punches are usually designed to last for 10,000 – 100,000 strokes.

Another technique for the generation of via holes is the use of a computer-controlled laser. This procedure is being used in industries that require a wide variety of parts or parts with varied hole patterns. One equipment manufacturer[14] claims that with a single head system, via diameters from 50 to 350 μm (0.002 to 0.007 in.) can be laser drilled in tapes up to 1.016 mm (0.040 in.) thick at a rate of 30 holes per second. With this system it is very easy to reprogram the computer to generate any hole pattern desired. Care must be taken to assist the removal of the debris generated so that it does not recondense on the surface of the tape-cast material.

After the tapes have been blanked to size and shape and the hole patterns have been established, the cards as they are now called, are ready for the next step in the manufacturing sequence. In some cases that step is the sintering of the part, but in most operations the next step is metallization and lamination to form multilayered structures. The lamination process will be described in the next section, since that is a process that is distinctly connected with the tape casting process. Metallization will not be covered, since that has been the subject of several books in the past.

5.3 CALENDERING AND LAMINATION

Two procedures that involve the application of pressure to either a single layer of green tape or to a stack of layers of green tape are calendering and lamination. These procedures will be reviewed in this section.

5.3.1 Calendering

By definition, calendering is the application of high pressure to a flat sheet of material by passing it through two or more rollers that revolve against each other. The process is commonly used in the paper industry to produce glossy and smooth finishes. The calendering process is better known in the ceramics industry as an alternative process to tape casting: roll compaction. In that process, the feed material is a spherical spray-dried powder that is compacted by pressing the particles together between two large counter-rotating rolls at a relatively high pressure. The resulting tape is very much like its tape-cast cousin. It is flat, smooth, and flexible and can be punched into shapes or can be rolled at the end of the processing equipment. This is not the process we will be describing in this section, since that is a subject for another book on processing. The calendering that we refer to here involves passing a tape-cast sheet, usually on a carrier, between two rolls that are set at a predetermined gap to squeeze or compact the green tape at a high pressure to a final thickness. This can be done with the edges free to move laterally toward the edges of the rolls or with the edges of the tape constrained from lateral movement by a rail or channel of some sort on one of the rolls. There are several reasons for calendering a green tape:

- A higher green density can be achieved, which lowers the sintering shrinkage.
- Some or all of the differential shrinkage between the x and y directions can be eliminated.
- A smooth surface finish can be imparted to the green tape surfaces.
- Much of the warping caused by one-side drying during tape casting can be eliminated.
- More uniform thickness can be achieved both across and along the tape.

Several manufacturers have included this processing step in their tape-processing lines for one or more of these reasons. The use of calendering is not widespread in the industry, however, and very little has been disclosed about its use.

5.3.2 Lamination

The basis for the production of MLC and MLCC (multilayered ceramics and multilayered ceramic capacitors) is the ability to laminate together several layers of green tape to form a structure that sinters together into a monolithic solid ceramic part. This procedure for the lamination of green tape cast-sheets was invented in 1965 by W. J. Gyurk of RCA.[15]

All of the basic parameters that must be controlled for lamination were described in that patent. These are temperature, pressure, and time. The temperature for lamination is dependent upon the organic phases present in the tapes, that is, the binder and plasticizers. The type of binder and the final glass transition temperature (T_g) produced through the plasticizer addition will determine the maximum and minimum lamination temperature that can be used. The number of layers and the final thickness of the tape stack to be laminated must also be taken into consideration, since the time at temperature must be sufficient to heat the entire stack as well as any parting layers that may be used to prevent sticking to the heated pressure platens. Sufficient pressure must be applied to provide intimate contact between the layers without distorting the laminated part and thereby changing the final dimensions. This distortion can be eliminated by using a closed die assembly instead of an open-sided system. A typical lamination procedure is as follows:

- Stack the metallized parts in the exact order desired using alignment pins. Make sure that the tapes are stacked top surface to bottom surface, since there is some binder segregation in most tape-cast products, and this will prevent binder-poor side to binder-poor side bonding, which can cause delamination.

- Use a parting substrate such as silicone-coated Mylar on top and under the stack to prevent sticking. Usually 75 μm (0.003 in.) or less in thickness is used.

- Apply a low pressure and allow the stack to preheat for a long enough time to reach equilibrium. This can range from seconds to minutes, depending upon the thickness of the stack. Standard temperatures used for most of the organic systems in tape-cast products range from room temperature to 100°C.

- Apply the full lamination pressure. The typical pressures used range from 1.4 to 138 MPa (200 – 20,000 psi).

- Hold the pressure for times ranging from seconds to minutes. Usually about 3 to 4 minutes is adequate for full lamination.

The art of lamination for the production of MLC and MLCC packages has evolved through the years, and several innovations have been added to improve the process. One of the most common problems observed with heated platen lamination has been the application of uneven pressure to the stack due to slight misalignment of the platens. It is very difficult to align and keep aligned platens that are subjected to high heat and pressure. This can also be caused by slight differences in tape thickness that add up to a large differential over several hundred layers. Uneven pressure application during lamination causes distortion during sintering due to uneven shrinkage of the part. It also is a source for interlaminar cracking. One of the "tricks" used to minimize these effects is to use multiple laminations of the same stack with rotation of the stack between pressings. Another common practice is to punch the alternate layers in a stack using a 90° rotation, that is, one layer would be punched with the tape in the casting direction and the next layer would be punched with the tape in the transverse or cross-cast direction. This actually provides two benefits:

(1) Any thickness differences in the tape cards would be rotated and thereby averaged out, and

(2) any difference in x and y shrinkage would also be averaged and would provide a laminated part with fairly uniform shrinkage in both directions.

A much better approach to the elimination of uneven pressure application is the use of a relatively new technique: isostatic lamination.[16] In this process the layers are stacked on a fixture plate and sealed in a vacuum bag. The bag (or several bags) are then immersed in a pressure chamber that is filled with heated water or a mixture of water and glycerin. Most isostatic pressure chambers operate between 34.5 and 68.9 MPa (5000 and 10,000 psi). Typical temperatures range from 70 to 90°C and the time for the complete cycle is from 3 to 10 minutes. The manufacturers state the following advantages of isostatic lamination:[17]

- Even distribution of lamination pressure, which prevents distortion upon sintering.
- Shrinkage is precisely controlled, and part-to-part variation is eliminated.
- Excellent heat transfer to all areas of the part ensures uniform lamination.
- Multiple packages can be processed in a single cycle.
- No rounding of edges or camber.
- No edges have to be removed—the entire part can be used.
- No multiple lamination with rotation has to be performed.

Another technique that has been described recently is the use of an adhesive for lamination.[18] In this process, the lamination is accomplished by gluing at room temperature and at very low pressures using an adhesive tape. The authors report that homogeneous monolithic sintered structures resulted using this technique.

In the manufacture of MLCCs today, as many as 300 layers are being laminated to produce a monolithic structure with excellent electrical properties. Lamination is definitely one of the most important downstream processing steps in the manufacture of tape-cast products today.

In the next chapter we will review some of the many applications of tape casting for the manufacture of a wide variety of products. The list is not meant to be all-inclusive. It is meant to show the diverse nature of the products and where they are used.

CHAPTER 6

Applications of Tape Technology

As we mentioned in the Foreword, the uses for tape casting as a fabrication tool have advanced well beyond the scope envisioned by the pioneers in the field some 50 years ago. In this chapter we will review some of the standard applications as well as some uses that are only beginning to be recognized as potential applications for this unique fabrication technology. Almost everyone involved with ceramic processing knows that thin, flat sheets can be produced using this technique, but how many ceramists appreciate the fact that 1.2 × 2.5 m (4 × 8 ft.) sheets that are 0.5 to 0.76 mm (0.020 to 0.030 in.) thick are produced by tape casting for use in the manufacture of molten carbonate fuel cells? These are some of the diverse applications that we will discuss in the sections that follow.

6.1 SUBSTRATES

For many years almost all of the substrate materials produced for the electronics industry were manufactured by tape casting. Substrates can be defined as the carrier or "backbone" of the electronic circuit. They are the ceramic insulator upon which the circuitry is deposited and patterned. These substrates ranged in size from as small as 6 × 6 mm (0.25 × 0.25 in.) to as large as 30 × 30 cm (12 × 12 in.), and in some cases larger sizes were produced. What all of these substrates had in common was their very small thickness, usually 1.5 mm (0.060 in.) or less. Thin-film substrates, those that carry either sputtered, evaporated, or chemical vapor–deposited circuitry on their surface, are almost all manufactured by tape casting. The standard for thin-film substrates is a 99.5% aluminum oxide material produced using starting powder with a very fine particle size. This combination yields a sintered substrate that has a surface microstructure with a grain size of 1 μm or less and an as-fired surface finish of less than 0.075 μm (< 0.000003 in.) CLA (center line average). This very smooth surface is required, along with a very low concentration of surface defects such as burrs and scratches to produce the very fine-line circuits in use today. In recent years beryllium oxide and aluminum nitride have been produced by tape casting for use as electronic substrates.

Today most thick-film alumina substrates are produced by roll compaction, which is a process more akin to powder pressing. A thick-film substrate is one that has screen-printed components on its surface. The substrate acts as the carrier for these components and also provides the electrical insulation to isolate the various components from one another. These components can be conductors, resistors, or capacitors, or any combination thereof. In the roll compaction process, the powders to be processed are spray-dried and then compacted into a flexible tape by squeezing the particles together between a pair of rolls that turn counter to one another. The diameter of the rolls and the spacing between the rolls determine the tape thickness. Tapes thicker than about 1.5 mm (0.060 in.) are usually processed by roll compaction. Some thick-film substrates are still manufactured by tape casting. The majority of the thick-film substrates are manufactured from a 92 to 96% alumina composition. The glass phase that is present in most of these substrates is essential for good adhesion of the screen-printed metallization to the surface and through the vias (interconnections from surface to surface of the substrate).

The substrates that we have been describing are single-layer substrates. There are a wide variety of multilayered packages that perform the same function as the single-layer substrate—that is, as a carrier for components that also provide insulation of one component from another, but which occupy a smaller area since some of the layers are internal to the package. The next section will discuss these multilayered ceramic packages in some detail, since they are manufactured primarily by tape casting.

6.2 MULTILAYERED CERAMIC PACKAGES

Multilayered ceramic packages (MLC) would not exist if tape casting had not been invented. The basis for the multilayer industry is the ability to individualize layers with respect to metallization and via interconnections and then to laminate a set of these individual layers together into a package that can be sintered into a monolithic structure. Multilayered ceramic packages with as few as two layers to structures with as many as a hundred or more layers are commonplace in the electronic ceramics industry today. The lamination process has already been described in the previous chapter. Some of the uses for multilayers will be reviewed here.

The first reported use for a multilayered ceramic package was published in 1961 by a group from RCA.[1] In this report, an application for the production of a quartz crystal oscillator was described. Since that time, ferrite memory and logic devices, tunnel diode packages, transistor packages, power diode packages, hybrid circuit modules, and small printed circuit boards have all been manufactured by the multilayer technology. The essence of this technology is the ability to house the semiconductor device or devices in a strong, thermally stable, and hermetic environment. In today's technology these packages take many forms, ranging from dual in-line packages to chip carriers and pin-grid arrays. The heart of IBM's computer technology is centered around the ability to produce these multilayered packages. It has been stated that IBM has invested over $1 billion in the development of multilayer ceramic technology and in the follow-up capitalization to manufacture these MLCs.[2]

The major advantages of this technology include the ability to provide high-density interconnections with relatively short path lengths, which lead to higher-speed computers; excellent thermal dissipation; the ability to operate at elevated temperatures, excellent structural strength; provision for a hermetically sealed transistor enclosure, excellent dimensional stability, and an overall low thermal coefficient of expansion.

The original multilayered ceramic packages were developed around what is today known as a high-temperature cofired ceramic (HTCC). In this technology the ceramic was primarily an alumina-based material that ranged from 85 wt% to 99 wt% aluminum oxide. The bulk of the packages produced fall in the 90 to 96 wt% range, with the remaining 4 to 10 wt% being composed of glass-former materials such as clay and talc. IBM actually used glass frits in their HTCC packages at about a 10 wt% concentration.[3] The typical thick-film metallization for these high-alumina tapes is tungsten, molybdenum, or molybdenum-manganese. These are classified as refractory metals, since their melting points are extremely high. The high melting points are required for the co-sintering process with the alumina packaging materials. Both the conductor paths and the via interconnections are prepared by conventional thick-film techniques such as screen printing and/or stencil printing. The ceramic layers in most multilayers are about 0.127 to 0.254 mm (0.005 to 0.010 in.) thick. The via holes that connect the layers are in the same range with respect to their diameter. After the layers are metallized and the stack is laminated, the package is cofired at temperatures ranging from 1500 to 1800°C. The exact temperature is a function of the ceramic formulation. Since the metallization is a material that is easily oxidized at these temperatures, the use of a reducing atmosphere such as hydrogen is necessary. A binder-burnout cycle is usually built into the overall sintering time–temperature curve. This usually includes a hold somewhere between 300 and 600°C. Time periods of 18 hours or more for binder removal and sintering are not uncommon in the manufacture of the HTCC packages.

A subset of the MLC technology that has been introduced in the past 10 to 15 years is the low-temperature cofired ceramic (LTCC) package. These packages have been developed since they can utilize gold, silver, or copper conductor materials, which provide much better electrical conductivity for the interconnect circuitry. Once again, this results in faster computer operation, since the chips can communicate with each other much faster. The LTCC packages are made of materials that sinter at temperatures lower than 1000°C. The packages are also composed of materials that have a lower dielectric constant than does the HTCC alumina-based package. The combination of a lower dielectric constant and the higher electrical conductivity of the metallization is desirable in the higher-frequency applications of today.

The typical material used in the LTCC packages is a glass-ceramic base. Some of these materials are actual glass-ceramics in that they are formed from a glass frit and then are recrystallized into a polycrystalline ceramic by using the proper heat treatment cycle. Other materials are prepared by mixing a glass phase with a ceramic filler phase. Typically the glasses used are borosilicate and alkaline earth aluminosilicate compositions, but a 96% silica glass such as Vycor™ has been used by some manufacturers as a part of the glass composition.[4] The goal, in the addition of all of these glass components, has been to yield a low dielectric constant and low thermal expansion coefficient (TCE). The low TCE is important in order to match the TCE of the silicon and GaAs chips that are to be attached. The typical filler materials used in these glasses are alumina, mullite, cordierite, and silica. By judicious selection of the glass(es) and filler(s) the properties of the resultant material can be "tailored" to suit the need of the package.

6.3 MULTILAYERED CERAMIC CAPACITORS

The first multilayered ceramic capacitor was described in the original paper on tape casting by Howatt and his coworkers.[5] It could be said that tape casting was invented to produce the thin layers used in the production of multilayer capacitors. Howatt's exact words were, "It was also found possible to combine the firing of the ceramic pieces and the electrodes. The dry sheets were coated on both sides with a gold-platinum paste (Hanovia Chemical Co., No. 14) and stacked several squares high. These were fired successfully into one dense block in a single operation. Units of very large capacitance in small volumes were made by this method." Howatt had invented the multilayered capacitor! The original capacitors, before the use of tape-cast sheets of a dielectric material, were produced by interleaving thin sheets of natural mica with electrodes of tinfoil. By connecting alternate electrodes together, one could arrange the capacitor elements in parallel, and the capacitance values were additive. Thus, by varying the number of mica layers, a range of capacitance values could be generated. Stetson has described the evolution of the multilayered capacitor in some detail in his paper on multilayer ceramic technology.[6] In that paper he attributes the modern-day multilayered ceramic capacitor (MLCC) technology to Warren J. Gyurk. Mr. Gyurk did indeed advance the technology by inventing the registrated lamination process for the production of multilayered monolithic ceramic bodies.[7] There are other ways of producing multilayered capacitors, but the standard for today's production is tape casting.

In today's world, tape casting technology has been pushed very close to its limits by capacitor manufacturers who are constantly trying to make thinner and thinner tapes. Tapes less than 3 μm thick have been produced in the laboratory, and 5-μm tapes are being manufactured by some companies. The goal of this exercise is to "increase the volumetric efficiency of the MLCC,"[8] that is, to raise the capacitance of the device while reducing the area consumed on the circuit board. Some of the increases will be achieved by the development of unique and better dielectric materials. The rest will be generated by manufacturing improvements. The production of thinner tapes and the equipment to handle them during the metallization and lamination steps will be absolutely necessary.

Equipment is now available to produce the thin tapes (less than 3 μm). There is also equipment available to screen-print and laminate these very thin tapes. Multilayered capacitors with 5-μm layers that are 300 layers thick are currently being manufactured.[9] Maher[8] states that there is a practical limitation where the dielectric constant of the material itself reaches its peak value as a function of the sintered grain size. If the grain size is in the range of 0.8 to 1.1 μm then there is a limitation on the minimum tape thickness at about 4 to 5 μm in order to maintain the dielectric breakdown properties of the chip capacitor. Maher also contends that a minimum of about 5 grains in series is desirable for reliability. It is generally believed that the 3-μm limitation on tape thickness will be the norm in the future.

Future trends will include a further decrease in capacitor size to keep up with the miniaturization of end-use equipment. This will necessitate layer counts as high as 300 to provide the capacitance values. Another trend is a move away from precious metal electrodes such as palladium to base metals such as nickel or copper in order to reduce the cost of the chip capacitors.

6.4 OTHER ELECTRICAL AND ELECTRONIC COMPONENTS

Along with the multilayered ceramic capacitors, there are many other categories of electronic ceramic materials that are active by nature—that is, they perform a function other than that of a simple insulator in the electronic circuitry. Some are conductors or semiconductors, while others act as resistors, sensors, electrooptics, or magnetic components. In this section we will describe some of the electronic devices that depend upon tape casting as a forming technique.

In addition to the multilayered ceramic capacitors just described, many of the barium titanate-based compounds that exhibit high dielectric constants are used in single-layer tape-cast capacitor devices. Relaxor materials such as lead magnesium niobate (PMN), which are characterized by high dielectric constants, broad dielectric maxima, and low sintering temperatures, have been manufactured in thin sheets by tape casting.

Thin sheets of piezoelectric materials are used in sensors, buzzers, and actuators. In addition to the conventional vibrators, pressure and acceleration sensors are now also being manufactured from these materials. Lead zirconate titanate (PZT) is one of the most common materials used for these applications. The trend is to produce thinner and thinner and smaller and smaller parts. Therefore tape casting has become the manufacturing route of choice. One of the basic applications of piezoelectric ceramics is as a gas igniter where a spark is generated by the piezoelectric under an applied mechanical stress. Microphone discs are also prepared from thin sheets of piezoelectrics that are configured as bimorphs, that is, two thin sheets of a material such as PZT bonded with a metal plate. The mechanical vibration caused by sound waves can be translated into electric impulses by this type of device. By reversing this process, that is, by applying voltage to a device such as this, a buzzer can be generated. Thus the piezoelectric effect: electrical energy is changed into mechanical energy and vice versa.

Actuators are among the latest applications for thin sheets of piezoelectric materials. Actuators are small displacement elements for the precision positioning of optical and other motors and machinery. All of these piezoelectric devices operate on the principle of electric field–induced strain.

A special class of materials that is also ferroelectric are electrooptic ceramics. Materials such as lanthanum-modified lead zirconate titanate (PLZT) produce excellent electrooptic devices. These polycrystalline ceramics exhibit voltage-variable behavior—that is, they can be switched from optically transparent to opaque by the application of voltage. Most of these devices, which are used for shutters, modulators, and displays, are processed by hot pressing to full density. Experiments in many laboratories are being carried out to tape-cast these materials into thin sheets. The main problem encountered to date has been the ability to sinter to full density. The use of nano-sized powders has helped in this regard. The ability to tape-cast large sheets could open a wide variety of applications for these materials.

Many different types of sensors have also been fabricated by tape casting. One of the most widely used sensors is the oxygen sensor used in automotive and other applications. These sensors are based upon solid-state conductivity of the ceramic elements at elevated temperatures. Most are fabricated from stabilized zirconia, which is the electrolyte in the device. The zirconia is a solid ionic conductor that transforms oxygen partial pressure or activity gradients into an electrical signal, which can then be further processed to perform control activities such as carburation or ignition modification. The tape-cast sensor is much smaller than the conventional oxygen sensor and has a much more rapid response time.

Since thin layers can readily be manufactured by tape casting, it has been used for a wide variety of sensor applications. Most of these sensors are based on the semiconducting ceramic materials that are used to monitor and control environmental variables such as temperature, relative humidity, and a wide variety of gases. Negative temperature coefficient (NTC) thermistors are typically fabricated from $NiMn_2O_4$. These materials have a very stable temperature coefficient of electric resistivity. They are used in devices that precisely measure and control temperature as well as those that provide temperature compensation for other nonlinear components and circuits. Positive temperature coefficient (PTC) resistors are manufactured from doped $BaTiO_3$. The PTC corresponds to a significant increase in resistivity with temperature. The principal applications for the PTC devices are self-regulating heaters, current surge protection, overcurrent protection, and temperature sensors. Gas sensors are produced from a variety of different oxides such as SnO_2, ZnO, and $MgCr_2O_4$. These ceramics are porous, allowing free access to the air, and the resistance is modulated by selective adsorption of atmospheric ingredients such as humidity (in the case of a humidity sensor). Applications include controls for microwave ovens, low-cost humidity controls, and alarms for dangerous gas buildup.

The multilayered chip inductor is a special use of ferrites. Since these parts are ultra miniature in size and light in weight, they can be manufactured using tape casting and lamination. These devices are surface-mounted on circuit boards and are used in high frequency applications such as mobile telecommunications equipment. They are considered to be very reliable, since they have the monolithic construction of a multilayered ceramic. As higher-and higher-frequency applications are generated, other materials such as the high-silica, low-dielectric-constant components used in LTCC packages will be used as multilayered chip inductors.

Varistors, or voltage-dependent resistors, have also been manufactured using tape casting. Some of the surge protection devices, such as SiC parts, have been manufactured using this process, since thin sheets are easily formed and the disk-shaped parts can be punched using "cookie-cutter" tooling. Although not currently the manufacturing process of choice, we believe that tape casting will ultimately be used for the production of many of these thin, large-area devices.

There are many references in the literature about the use of tape casting for the manufacture of high-temperature "superconductors." The process permits the fabrication of thin sheets that can be very long. Attempts have been made to produce flat wires using this technique, especially when the wires are wound onto a central core of some sort. It is not known whether this technology or the use of tape casting for the fabrication of thin shapes of this interesting material will ever be viable, but much research is being done in that direction.

6.5 FUEL CELLS FOR POWER GENERATION

Since thin layers of materials are a necessary part of most fuel cell construction, tape casting has been used extensively in this field. There are two basic types of fuel cells: solid oxide fuel cells (SOFC) and molten carbonate fuel cells (MCFC).

6.5.1 Solid Oxide Fuel Cells

Solid oxide fuel cells are generally based on the same principle as the oxygen sensor described previously. That is, electrical energy is produced from a reaction of gases such as hydrogen and oxygen or natural gas and oxygen with water as a by-product. Many of these stabilized zirconia-based fuel cells have been manufactured for use in nuclear submarines and in spacecraft of all types. The electrolyte in these fuel cells is the stabilized zirconia, which becomes a conductor of oxygen ions at elevated temperatures. In many cases the zirconia membrane, which is relatively large in the x and y directions and has a very thin crosssection, is manufactured by tape casting. All of the so-called planar cell configurations use the tape casting process. Some of the power generation applications will produce as much as 250 kW, which may replace a gas turbine combustor.

6.5.2 Molten Carbonate Fuel Cells

Molten carbonate fuel cells (MCFC) use an electrolyte made of lithium and potassium carbonates. They operate at 650°C and produce fuel-to-electricity efficiencies approaching 60%. The molten carbonate fuel cell generates power by the reaction of natural gas and oxygen to produce electrons and water as a by-product. All of the components in the MCFC are manufactured by tape casting. The electrodes, both the anode and the cathode, are tape-cast and sintered, and the electrolyte is simply cast. The lithium–potassium carbonate is used in the as-cast condition and is mounted along with a lithium aluminate separator between the electrodes inside of a closed container. The binder/plasticizer is burned out of the electrolyte during the start-up phase of the operation, and the electrolyte melts and fills the porosity in the lithium aluminate skeleton matrix. Several of these MCFCs are in operation at least on a pilot-plant scale. There is a 2-MW unit in Santa Clara, CA that is tied into their municipal power grid, and there is a 250-kW unit at the naval air station in San Diego, CA. It is anticipated that several of these power generation stations will come on-line during the next decade. Cells as large as 1.2 × 2.4 m (4 × 8 ft.) are projected for manufacture during that time frame. This is amazing, considering the fact that the thickness of these tapes is on the order of 0.5 mm (0.020 in.)!

An interesting application of an old process invented by one of the authors[10] has been reviewed recently by Japanese authors at the Matsushita Electric Industrial Co., Ltd.[11] In their paper they describe a process for fabrication of the electrode and electrolyte at the same time, using a multiple casting technique in which one layer is cast on top of the other. The authors claim a simplified manufacturing process as well as better contact between the electrodes and the electrolyte. Once again, the use of tape casting was an important aspect of the processing.

6.6 FUNCTIONALLY GRADIENT MATERIALS

A considerable amount of research is being conducted to produce functionally gradient materials (FGMs) for a wide variety of applications. The ability to tape-cast and laminate several layers of materials with differing chemical compositions makes these FGMs possible. We will describe a few of these that have appeared in the literature and will add some speculative ideas of our own.

6.6.1 Metal/Ceramic Composites

In the processing of metal/ceramic composites there are often severe problems associated with the significant differences in coefficients of thermal expansion (CTEs). Stresses that develop between the ceramic and metallic layers can often lead to delamination failures. In order to minimize this differential at the ceramic/metal interfaces, functionally gradient materials have been developed. By using a series of layers, each with a slightly different ceramic/metal composition, and then by laminating these layers together, a very gradual transition can be made from the "pure" ceramic to the "pure" metal layer. Tape casting is an excellent technique for forming this multilayered structure, since layers can be prepared with essentially the same organics, and thus the lamination and burnout characteristics are the same. Tape-cast layers can be made to any thickness and are very easy to process into the FGM composite. A process based upon this technique has been described by Sabljic and Wilkinson.[12]

6.6.2 Electronically Graded Composites

Another very interesting use of FGM technology was recently described in a U.S. Patent.[13] In this work a laminated structure of ceramic ferroelectric materials with adjacent layers of barium strontium titanate and other oxides is stacked in order of descending oxide content and sized in thickness to produce a generally equal capacitance across each layer, resulting in a material having a graded dielectric constant for use in a phased array antenna. The layers are produced by tape casting, and the structure is fabricated by multilayer lamination.

We have only reviewed two possibilities for this intriguing new technology, which is based upon the ability to cast thin layers and to laminate them together to form a monolithic structure with entirely different properties on the two outside surfaces. One can envision other uses for the FGMs, such as the ability to build in metallic heat sinks or ground planes inside a ceramic insulator package (or on one surface). It is believed that design engineers will use this approach to solve problems when the technology becomes better known throughout the materials processing industry.

6.7 SEPARATORS FOR BATTERIES

Battery separators must have very thin crosssections in the z or thickness direction, and they usually have fairly large x and y dimensions. A separator is a nonconducting, corrosion-resistant material that keeps the electrodes from touching and shorting out the battery. This once again makes tape casting a production method of choice. The separators commonly used in lithium metal sulfide batteries are AlN, MgO, and BN. The requirements for these materials are high purity, thin cross-sectional area, mechanical strength, and corrosion resistance to the chemicals used. There are a number of materials suppliers and battery manufacturers who are involved in tape casting to produce these separator materials.[14]

6.8 STRUCTURE-CONTROLLED MATERIALS

It has been known for many years that the tape casting process, as a result of the shearing action under the doctor blade, tends to create a preferred orientation of the grains in the tape.[15] With some materials the effect is small, but with others the properties of the sintered part are dramatically changed. Researchers in Japan[16] have developed a high-performance silicon nitride ceramic as a result of this in-process structural control. They describe a procedure in which highly pure beta-silicon nitride particles, 4 μm long and 1 μm in diameter, are incorporated into a slurry formed by alpha-silicon nitride powders. The slurry is then tape-cast to produce green sheets 100 μm thick, which are stacked and pressed into laminated sheets, then sintered at 1850°C in a nitrogen atmosphere. During the forming process, the seed particles are aligned within the cast plane so that the prismatic particles that grow from the seed crystals form a uniform planar dispersion within the matrix. This control of the cast and sintered morphology is the basis for the excellent mechanical properties that result. The researchers report a Weibull constant as high as 50, compared with a value of 20 for conventionally processed material. This process, which is currently being called templated grain growth, is being used to improve the mechanical properties of ceramics and ceramic composites in many universities today.[17,18]

6.9 RAPID THREE-DIMENSIONAL PROTOTYPING

Rapid prototyping (also known as solid free-form fabrication) was originally introduced by Lone Peak Engineering and Helisys, Inc. in 1992.[19] This technology depends upon tape casting to produce the individual layers, which are then cut and laminated together to produce a three-dimensional part. Rapid prototyping processes combine computer-aided design (CAD) with this layering process to fabricate the part. When the conceptual design has been finalized, a computer-generated model is electronically sectioned into layers of a predetermined thickness. When the layers are reconstituted by lamination, the original part is reconstructed. Information about each section is transmitted to the prototyping machine, and the part is then built by the machine, layer by layer. The thinner each individual layer is, the more detailed the final shape becomes. In the Lone Peak process, the individual layers are cut by a computer-directed laser. After the layers are laminated together to form the final shape, the part is then subjected to a binder burnout process and a final sintering process. Very complex shapes have been formed by this technique, including parts with internal chambers.

The beauty of the RP technique is the rapid turnaround time from design to a final model, a process that manufacturers claim can be accomplished in approximately four days. Researchers at the University of Dayton[20] are using rapid prototyping to produce fiber-reinforced ceramic matrix composites for use in armor and electrical components. A very exciting application of this technology is being done by Zimmer, Inc., Midwest Orthopedics, and the U.S. Department of Energy's Argonne National Laboratory. These laboratories are working on a joint project to speed the production of joint and bone-segment replacements.[21] They claim that the rapid prototyping process will permit surgeons to create a replacement bone segment for a patient by translating X-ray information into a digitized computer program, which can then be used to build the ceramic prosthesis. The materials development program to control the porosity of the ceramic to imitate real bone is a very crucial part of this program. This is a very new and dynamic use of tape casting technology and one that was never conceived of when the process was developed to produce thin two-dimensional parts.

6.10 LITHIUM ION BATTERIES (POLYMER BATTERIES)

Much tape casting equipment has been purchased in recent years by the manufacturers of lithium ion batteries in order to produce the components that go into these products. Lithium ion battery usage has soared with the proliferation of cell phones and laptop computers. The lithium ion batteries are rechargeable and very lightweight, and they are much safer than the batteries that preceded them in these portable applications. The specific application of tape casting to the manufacture of these batteries resides in the ability to cast thin films of the lithium ion–containing polymer onto a carrier such as Mylar or other polymeric material. This electrolyte layer can then be laminated between the anode and cathode layers, which are tape-cast onto aluminum foil and copper foil. The multilayered structure thus formed can be rolled up into a cylindrical shape or can be used flat. This technology is in its infancy and, at this point, depends heavily upon tape casting.

6.11 BRAZING ALLOYS

For many brazing applications there is a need for a thin, essentially two-dimensional tape that can be fitted in place using an adhesive.

Tape-cast brazing alloys are flexible, and when laminated to an adhesive layer they can be applied to the surfaces to be brazed even in very complex and convoluted areas. The metallic powders are mixed with the organic components and the slurry is tape-cast in a very conventional manner. The thickness is controlled by the setting of the doctor blade. A wide variety of thicknesses are used for this application, although most of the tapes are in the 0.254 to 1 mm thickness range.

6.12 THIN SHEET INTERMETALLIC COMPOUNDS

Many of the intermetallic compounds are very difficult to process into thin sheets because they exhibit very low ductility at ambient temperatures. Powder metallurgical techniques coupled with tape casting have been employed to process these materials into continuous thin sheets which, after removal of the polymer binder and plasticizer, can be sintered and rolled to full density with a sequence of cold work and annealing steps. The use of tape casting has recently been reported[22] for the production of iron aluminide (FeAl) thin sheets with excellent properties. We believe that this processing technology can be used for other intermetallic compounds, especially for the production of continuous thin sheets.

We realize that our review of tape casting applications is not all-inclusive, and it was not intended to be so, but we have tried to provide a broad-brush perspective of the wide variety of uses for this still-evolving fabrication technology. The range of uses in the future will be dictated by the ingenuity of the design engineers who become familiar with the process, and hopefully this book will provide some impetus in that direction. Any material, whether it is metallic, ceramic, or polymeric, is a potential candidate as long as a thin, essentially two-dimensional tape is part of the structure. We say "part of the structure" since, as we have described previously, three-dimensional structures are now being generated using these two-dimensional tapes.

CHAPTER 7 Water-Based (Aqueous) Processing

WATER-BASED (AQUEOUS) PROCESSING
Swan dive, belly flop, or mile swim?

Oswald Chambers is generally the person attributed with the "slippery slope" picture. When you're on the pinnacle (where you should be), one step in any wrong direction leads to a drop. One wrong step is easy to make, but getting back to where you should be is a hard and frustrating task. What's a famous preacher got to do with it? Tape casting started as a water-based production method. It fell way, way down that slippery slope. It's a long way back up.

One of the hardest parts of water-based slip development is that everyone in the capitalist industry wants it. So when manufacturers find something really great, they hide it instead of telling others. Don't worry...we don't think capitalism is such a horrible thing, or that the companies keeping secrets are evil, but trade secrets don't aid technology development in this case, now do they?

One of the most troublesome aspects of water-based slip development is the high dielectric constant of the "universal solvent," which, after a few steps of logic, which we'll exclude to give you a puzzle to ponder, leads to the point that water-based slip formulations tend to be aggressively powder dependent. Where a PVB slip might be useful for alumina, zirconia, titania, ferrite, and nickel metal, a water-based slip will tend to work only for one certain powder chemistry and fail miserably for other similar powders.

It's hard to believe, with all of the articles and patents printed or granted in the last 10 years regarding the switch to water-based tape casting, that tape casting originated as an aqueous processing method. Howatt[1] taught the use of porous plaster slabs in a type of "continuous slip casting" of a water-based suspension. Park[2] discovered the practical advantages of higher-volatility liquids, including the increased ability to dissolve film-forming binders, and we've been trying to go back to aqueous-only processing ever since. The increasing costs of purchasing, handling, recycling, disposing of, and protecting workers from organic solvents have been a tremendous source of motivation to explore environmentally benign (read "green") processing methods for tape-cast products. Disposal of EPA-defined hazardous air pollutants, operating within OSHA safety limits, obeying local and state fire codes for manufacturing equipment, and concern for adverse health effects due to prolonged exposure to certain organic solvents have all acted as motivating factors for research into aqueous tape casting systems.

Perhaps the main thing that hinders a mass migration to the "green" side of the manufacturing fence was mentioned in the first paragraph. Summing it up in a single sentence, "Is it practical?" Are the research dollars and additional processing expenditures going to pay off in the long run? The technology for water-based tape casting has existed for decades, improving in both expertise and component capability as time progresses, but is still not economically *practical* for many tape producers and manufacturers. In a head-to-head comparison between organic solvent–based tapes and water-based tapes, organic solvents are easier to process and make higher-quality tapes with superior performance whose green properties are easier to tailor to the specific needs of a particular manufacturing process. But if the water-based process is easy *enough*, and the tape is good *enough*, and strong *enough*, and repeatable *enough*....

This chapter is devoted solely to water-based tape casting. Any mention of the word *solvent* in this chapter refers to the "universal solvent," water, unless specifically stated otherwise. The ability of water to "dissolve" any and all materials (to some degree) is perhaps its greatest strength and greatest weakness. Water can dissolve (leach) ions out of the ceramic powders suspended in it; disassociate into acidic and basic species ($-OH^-$ and $-H^+$) and from there alter the surface chemistry of the suspended powder(s); and can form longer-range electrostatic forces between particles than most organic liquids that either attract or separate the particles (DLVO theory). It also has an exceedingly high surface energy (surface tension) compared to ethanol, methyl ethyl ketone, acetone, or toluene. These factors tend to make water-based processing quite a bit more challenging than processing in organic liquids.

The models developed in previous chapters, however, are general tools based on the structure and resulting properties of the tape and they may still be used with aqueous processing. These models include the mosaic tile model, Type I and Type II plasticizers, and, most importantly of all, Grandma's Cooking Pot. Since the health and safety benefits of water-based processing are obvious, the following discussion will focus mainly on the drawbacks of water-based processing.

Two basic approaches have been used in water-based tape casting. These approaches differ in the types of binder selected. In organic systems, all binders used were soluble in at least one of the vehicles chosen. This same approach in the water-based area leads to the use of water-soluble film formers, including certain celluloses, vinyls, and acids. The second approach is to stick with the polymer families we know well (acrylates and vinyls) and to form water-based emulsions of these resins. A caution must be issued at this point due to the fact that many in the field consider emulsions to be water soluble.

An emulsion, according to its basic definition, is a suspension of one material in a liquid in which it is *not* soluble. The emulsion as a whole may be considered "soluble" in another liquid, but in the case of binders for aqueous tape casting, the line of distinction must be drawn between water-soluble binders and emulsion binders. Specifically, a water-soluble binder is a polymer that *dissolves* in water, and an emulsion binder is a collection of insoluble polymer particles, "droplets," *suspended* in water. As pointed out by Doreau et al.,[3] the addition of emulsion binder to a slip is accompanied by an addition of water which is also in that emulsion. An 80 g addition of a 50% active emulsion only puts 40 g of binder (polymer) into the tape. Soluble binders may be added 100% active. The two binder families will be discussed separately in this chapter. Phenomena that are shared by both types of binder will be discussed later.

The dividing line between the two branches of water-based development is drawn between types of binders. The dividing line between organic *processing* and aqueous *processing* is also drawn at the binder. The dispersion of ceramic and/or other particles in water is much the same in water as in organic vehicles. The only additional variable is a much stronger dependence on pH. Aqueous dispersion of particulates has been covered ad infinitum in other texts and will not be repeated here. The mechanisms of deagglomeration, deflocculation and dispersion for varieties of particulate species, various dispersants, deflocculants, wetting agents, defoaming agents, and so on can be found throughout ceramic publications and summarized in various books.[4,5] Dispersion, deflocculation, and viscosity data must be collected for each system of interest and has been done, with tape casting specifically in mind, for a number of powder/dispersant/binder systems.[3,6,7,8]

Dispersing of the powders in the solvent (water) to form a stable dispersion is followed, as before, by the addition of binder(s) and plasticizer(s). The following sections will address the different types of binders separately along with common hurdles and possible techniques for overcoming these hurdles. By far, the major hurdles (often used synonomously with aqueous tape casting) are: foaming during milling, brittleness, foaming during de-airing, cracking, foaming during slip handling, dewetting, high slip viscosity, "curdling" and bubbles in the tape.

7.1 WATER-SOLUBLE BINDERS

Film-forming polymers that are soluble in water have been used in the ceramic industry for many, many years. Those knowledgeable in dry pressing are probably familiar with polyvinyl alcohol. Those versed in extrusion are likely familiar with one or more cellulose ethers. These polymeric binders are widely used to impart strength and stiffness to pressed, extruded, and slip-cast parts. These binders have also been explored for use in tape casting. Slip formulations using water-soluble polymers such as PVA[9] and celluloses,[10,11,12,13,14] as well as polyacrylic acid[6] can be easily discovered in published literature.

Perhaps surprisingly, these binder choices can be treated in exactly the same manner as the soluble binders used in organic vehicle systems. In the presence of a Type I plasticizer (water), increasing the amount of Type II plasticizer lowers yield stress and raises percent strain.[14] With the same materials, increasing solids loading (less free binder) lowers yield stress and percent strain to failure.[14] Type II plasticizers have also been seen to double as release agents,[13] as was described earlier for organic solvent systems. The differences between organic solvent systems and water systems do not appear until the "big picture" is taken into account.

The "big picture" in this case focuses mainly on the ability to plasticize these water-soluble polymers in the desired manner. Many plasticizers have been used and reported for both PVA and cellulose binders, but the vast majority fall into the category of Type II plasticizers. Glycerin and various glycols have been reported, and seen by the authors, to be very effective lubricants in the water-soluble tape casting systems. The earlier section on plasticizers, however, pointed out that while a Type II plasticizer is desired in some cases, the mandatory plasticizing agent found in all tapes is the Type I plasticizer or "binder solvent." The requirements of a Type I plasticizer are solution of the binder polymer and low volatility. Water-soluble binders are plasticized (Type I) by *water*, which satisfies both of these criteria.

The plasticizing (Type I) behavior of water in PVA or cellulose tapes can be used in different ways. Since high–Type I plasticizer content on the tape surface promotes blocking, water can be used to "glue" tape layers together[15] or to control tape strength. An excellent study on aqueous casting of Al_2O_3 by Chartier et al.[13] shows that higher binder content, which should result in greater strain, results in lower strain to failure when Type I plasticizer (water) content is significantly reduced.

In manufacturing, fluctuations of water content in PVA and/or cellulose tapes can prove costly. Fluctuations in atmospheric humidity/relative humidity can cause large changes not only in laminatability and flexibility but also in tape strength, brittleness during blanking or via punching, and dimensional uniformity. Large changes in humidity and temperature can result in catastrophic changes in green substrate size, via spacing, and firing shrinkage.

Another potential problem in water-soluble binder systems is the achievement of easily processable viscosities.[7] Celluloses are used as thickeners in some industries (food, pharmaceutical), and they behave similarly in tape casting. PVA slips also tend toward higher viscosities, but not to the extent of celluloses. Reported slip recipes have addressed this thickening effect by using additional water to lower slip viscosities. This results in castable slips, but it also results in wet:dry ratios of 6.5:1[12] or 8:1,[10] or *z*-dimension shrinkage of 77%.[13] The additional water also results in comparatively low solids to solids plus solvent ratios. Some reported formulations display viscosities of: 37,500 mPa·s (cP) at 38.6 wt% solids,[10] >17,800 mPa·s (cP) at 43 wt%,[12] or solids loading as low as 17.3 wt%.[13] The role of pH on the viscosity of the multicomponent slip is not widely reported.

Preparing a repeatable, homogeneous slip becomes more difficult with these comparatively high slip viscosities. The solution and mixing of a high-viscosity binder added 100% active is not always possible in reasonable processing times (24-hour rolling mill). For this reason, as well as other solubility issues with 100% active PVA or cellulose powders, water-soluble binders are often predissolved in water and added to the slip as a liquid solution. Most cellulose solutions can be prepared simply by rolling in a jar without media for 24 hours. PVA solutions can be prepared in the same fashion, or by dispersing the PVA powder in hot water (around 95°C) and cooling the suspension by further additions of cold water. Additions of predissolved binder typically make the task of homogenous incorporation faster and much easier.

The last processing problem touched on in this section, but certainly not the last processing problem for these binders, is the removal of trapped air or foam. The combination of low solids loading, high viscosity, and a relatively high Type II plasticizer content (glycols especially) often results in large volumes of entrapped air. It is not uncommon during the development of water-soluble systems to find a mill with no actual liquid but instead a fine, stiff foam that a pastry chef would be proud of. This foam is usually very difficult to reduce by vacuum techniques, since it can increase in volume dramatically (500% or more) before bubbles start to break, reminiscent of Shanefield's green monster.[16] Additions of an external defoaming agent are typically used to break the bubbles, often in conjunction with stirring, vacuum de-airing, or slow rolling without milling media. Other de-airing methods we have either been exposed to or told of include sieving the slip (filtering), centrifuging, and vacuum ball milling.

After the tape is cast and dried, the ambient environmental conditions are crucial to the performance of the tape during downstream processing. As mentioned earlier, airborne water can affect not only strength, flexibility, and lamination but also dimensional stability and uniformity of the tape-cast layer. Water as a Type I plasticizer can change strength, lamination parameters, and other properties as mentioned earlier. Water as a solvent, however, can change the size and shape of the tape. With a high wet/dry ratio, it is clear that the presence of water in the tape structure changes the physical dimensions of the layer. Swelling and continued "curing" of the tape due to the presence of water over time can be detrimental to downstream processing. Fluctuating humidity can swell and shrink the tape over its storage life. Prolonged storage in high humidity, as might be seen in shipping, can continue to cure the polymer and reduce or rearrange the vol% air in the tape. If the tape is rolled after drying, compressive stress on the inner surface and tensile stress on the outer surface will redistribute binder, particles, and air during storage if the water content lowers the polymer yield stress sufficiently.

Changes in the tape structure affect local green bulk density and lead to fluctuations in firing shrinkage. An analogy to organic solvent-based tapes would be to store a B-98 poly(vinyl butyral) tape in a partially saturated ethanol or MEK atmosphere. Tight temperature and humidity control are almost mandatory when high-tolerance parts are to be manufactured in water-soluble tape casting systems. Emulsion binders, being water insoluble, do not have this extreme sensitivity to humidity.

7.2 WATER-EMULSION BINDERS

Types of emulsion binders include acrylics,[17] vinyls,[18,19] polyurethane,[20] and others. An emulsion binder is a dispersion of nonsoluble binder particles in water. The direct ramifications of this fact are quite profound. First, the binder is not a continuous phase in the slurry. The tape casting slip is a dispersion of both inorganic and organic (binder) particles. The binder does not become or act like a continuous phase until partway through the drying process. A side effect of this is that the binder is not likely to displace dispersants on the particle surface.[8] Second, the dispersed polymer cannot be modified by a Type I plasticizer after it is suspended without coalescing some or all of the suspended binder particles or "breaking the emulsion." All Type I plasticizer interactions must be complete prior to emulsifying the binder. Third, suspension of the inorganic particles must be done in such a way as to create a tolerable environment for the binder suspension. There is now an additional particulate population that needs to be kept dispersed, and what may be good for one may not be tolerable by the other.

Addressing dispersion stability first: The incorporation of an emulsion binder must be approached from the viewpoint of merging two suspensions, not as the incorporation of a soluble tape casting binder. The use of emulsion binders requires an entirely different mindset for tape casting, and the trap of Grandma's Cooking Pot must be avoided. Soluble binders are most often incorporated into the slip by rigorous mixing, ball milling, or vibratory milling with grinding media. The energy of such milling is often far too high to maintain the integrity of the polymer emulsion. Under such conditions, the emulsion particles or "droplets" may coalesce, resulting in a lumpy slip consistency resembling milk curd. This is therefore commonly referred to as "curdling" of a slip. Other effective and less risky ways of incorporating an emulsion binder include stirring by hand or overhead mixer, swirling or using a magnetic stir bar for very small batch sizes, or jar milling without milling media. Since binder/plasticizer chemical reactions are not needed, and solution of the binder is not required, homogeneity can usually be achieved in about half an hour. Since emulsions are usually negatively affected by rigorous milling, and the first stage of dispersion for fine powders is mechanical separation (see dispersants section); the use of emulsion binders forces the use of two-stage slip preparation in most cases.[3,21]

In addition to mechanical stability of the emulsion, chemical stability must also be maintained. Prepared emulsions are suspensions of fine polymer particles or "droplets" that typically rely on a dispersant to stabilize the suspension. If the dispersant used is functional only in a limited pH range, that pH range must be maintained in the slip both before and after emulsion incorporation. This pH sensitivity of some emulsions may require careful selection of dispersant/deflocculant candidates for the inorganic material as well. While alumina may be easily dispersed in an acidic environment with no other dispersing agent needed, the emulsion of choice may only be stable in the pH range of 7 to 10. A dispersing agent would be needed in this system that functioned well in an alkaline slurry.

The area of chemical stability can also require passivation of the inorganic particle surfaces. Water is often called the "universal solvent" because it dissolves just about everything to some degree. The ability of water to dissolve various ions and compounds can be detrimental to the preparation of an aqueous-emulsion tape casting slip. In addition to pH sensitivity, some dispersing or stabilizing agents are sensitive to specific ions in solution.[8] In particular, the stabilizing agents in some members of the Duramax® family of binder emulsions display sensitivity to barium ions in solution.[22] While the emulsion may be stable in weak hydrochloric acid at pH 5, the acidic environment leaches barium from $BaTiO_3$ particles and creates a barium ion concentration too high for the emulsion to remain stable. In this case a basic solution is necessary even though the emulsion is not directly harmed by slightly acidic pHs.

One effective approach to the chemical stability issues surrounding emulsions is a stability diagram. Similar to a phase-field diagram defining crystal structures, a plot of pH vs. solids loading, pH vs. temperature, pH vs. dispersant concentration, or pH vs. binder concentration[7] can be used to define emulsion stability limits for different slurry formulations. Emulsion stability is also affected by temperature changes, but with the gentle processing already required for mechanical stability of the emulsion, overheating of the slip is usually not a concern. Another, more direct approach to stabilizing an emulsion is the incorporation of an additive that stabilizes the emulsion droplets. One example of this is the use of Triton X-405 in the water-based emulsion $BaTiO_3$ batch shown in Appendix 1. An emulsion supplier may or may not offer advice on suitable additives.

The last major change in approach when using emulsion binders is in the de-airing step. Water-based tape casting systems are known for trapping air. Entrapped air in a water-based system can lead not only to pinholes in the tape, but can also be nuclei for fish-eye demarcations or larger-scale dewetting issues that affect a larger area of the cast layer. De-airing of organic solvent–based slips is most often accomplished by a vacuum de-airing process. The high surface energy of water (unmodified) and the very low volatility of water combine to make vacuum de-airing of emulsion slips a generally poor choice. In slurries with MEK or isopropanol, the solvent vehicle is the highest-volatility component of the slip. Pulling a low vacuum on these slips not only enlarges bubbles past their breaking point, but also evaporates some of the solvent vehicle. Pulling a vacuum on an aqueous-emulsion slip not only provides the driving force for bubbles to rise to the surface, but also drives evaporation of the most volatile species in the slip.

In slips that contain ammonia groups, either as dispersants[7,23] or pH modifiers, it is very possible that ammonia is among the most volatile species. Vacuum de-airing forms bubbles high in ammonia content and puts more bubbles into the slip than it removes. Another result of volatilizing the ammonia is a change in pH. We have repeatedly seen, with a number of different emulsion binders, that vacuum de-airing an emulsion results in curdling of the slip at the slip surface if not the entire system. Vacuum de-airing can be used to remove air from the powder suspension but should not be done to a slip after the addition of an emulsion. The emulsion should be added in such a manner as not to incorporate air back into the de-aired slurry. Filtering the slip through a fine mesh screen can also be an effective tool for removing air from the slip.

The methods of plasticizing an emulsion binder are quite simple. Since the addition of a Type I plasticizer is not possible, only Type II plasticizers remain. If additional Type I plasticizer is needed, a lower-T_g emulsion should be sought out, forcing a change of binder as well. If suspension stability allows, emulsions with two different T_gs can be used in conjunction, resulting in a net T_g somewhere between the two individual components.[21] This approach assumes mutual compatibility of the two emulsions and Type I plasticizers. Type II plasticizer can be externally added[7,23] and behaves in much the same fashion as described by the mosaic tile model. As with the water-soluble systems, Type II plasticizers tend to create foaming problems and thus should either be mixed into the slurry like the emulsions, or during ball milling followed by a de-airing step prior to emulsion addition. External defoaming additives may also be used during the de-airing process, but care should be taken so as to not include any ions or compounds that will negatively affect the final, fired part (e.g., silicone-based defoamers).

One other option that has recently been marketed to the tape casting industry is an externally cross-linkable emulsion binder. This binder works together with a secondary additive (polyoxyalkylene-diamine MW = 400) to increase the 3-D polymer network strength by increasing the strength of attachment from one emulsion droplet to those surrounding.[21]

7.3 SHARED CONCERNS

7.3.1 Dewetting

Ion solubility and foam are not the only adverse effects of water. The high surface energy of water, even after wetting agents are used, hinders the water-based slurry in carrier wetting. This resistance to wetting, or dewetting, concern has been much more prevalent in emulsion systems than in soluble systems. Dewetting from a carrier surface can manifest itself in a variety of ways, depending primarily on the balance of energies at the slurry/carrier/air interface. Since, as a general rule of thermodynamics, all systems strive for the lowest energy state, the ternary interface shapes itself to minimize the overall energy of the system. If the liquid/carrier interfacial energy is greater than the air/carrier interfacial energy, the slip will try to minimize the area of liquid/carrier contact. This phenomenon is known as dewetting. If the liquid/carrier interface is lower energy per unit area than the air/carrier interface, then the slip will try to cover or "wet" as much area as possible. This energy balance is discussed in depth in most physical chemistry books under "contact angle."[24]

Forces arising from surface and interfacial energies, however, are not the only forces active at the tape edges. Slip specific gravity, volume (or in this case thickness), and viscosity also play a role in determining whether the slip will dewet from the carrier, and if so, how far. While surface energies may define the fluid contact angle, raising the weight of slip to be moved will increase the energy necessary to move it. In the same manner, a heavier or higher specific gravity slip will be harder to move and thus dewet less. These two options, however, require changes in either tape thickness or slip specific gravity, which are not changed without many downstream repercussions. There is, however, a more tolerable choice. Slip viscosity was defined earlier as a fluid's resistance to flow, or a kind of "internal slip friction." Increasing slip viscosity will not directly affect the contact angle, but will decrease the dewetting distance by adding an internal retarding force to the dewetting.

When a high contact angle is defined by the system's interfacial energies, the extent of dewetting (dewetting distance) can be limited by formulating the slurry with less water. The resulting higher solids loading causes an increase in slip viscosity and specific gravity, thus limiting dewetting. The combination of high contact angle and low dewetting distance will often result in tapes with beaded edges, similar to "rails." As mentioned previously, slips containing dissolved polymers such as PVA or cellulose generally have higher viscosities than emulsion-containing slips. For this reason these higher-viscosity slips are not known for their dewetting problems to the extent that emulsion slips are.

The last, and typically easiest method for combating a dewetting problem between water and a specific carrier is to simply use a different carrier film. Various tape casting carriers were mentioned earlier in this book, and each has a different interfacial energy with water, as well as a different surface energy. While a water-based slip may dewet from silicone-coated Mylar or lecithin-coated steel, polypropylene may be quite suitable. Keep in mind, however, that in addition to wetting, the carrier used also has to meet certain criteria such as strength and flatness.

7.3.2 Drying

While slips with water-soluble binders dry in the same fashion as organic solvent–based slips, emulsions have somewhat different behaviors. Three minor differences exist between water-soluble-binder slips and organic solvent–based slips. First, water requires more energy for evaporation (higher heat of vaporization) than most organic solvents used. Second, there is typically more vehicle volume to evaporate in an aqueous system than in an organic solvent system, due to the viscosity issue previously discussed. Third, the Type I plasticizer and solvent vehicle have the same volatility in a water-soluble system (since water is playing both roles), and drying the tape thoroughly also makes it less flexible and slightly brittle.

Emulsion slips have drying characteristics somewhere between the two extremes. The Type I plasticizer is pre-reacted with the binder prior to emulsification (or a low-T_g polymer is used in the first place), removing the need for special treatment in this area. Additionally, the binder is suspended as particulates, which avoids skinning issues during the initial stages of drying. While in organic solvent systems the first drying stage should be done slowly enough to avoid skinning the cast, the initial drying of an emulsion slip can be done as quickly as desired short of boiling the water. As the concentration of binder and inorganic phase increases with the removal of water, a "gel structure" begins to form, resulting in a three-dimensional network. Upon the formation of this network, drying proceeds in much the same fashion as with an organic solvent–based slurry.

The freedom offered by an emulsion binder to dry quickly in the first stage is a great benefit. Energy, usually in the form of heat, can be put into the wet cast more quickly than the higher-volatility organic solvents would allow. This freedom can result in shorter drying times than even organic solvents can provide, depending of course upon the specific slip system and tape thickness. We have seen one 0.060-in.-thick water emulsion cast dry more than twice as quickly as a comparable MEK/ethanol system, since drying the organic solvent system more quickly resulted in cracking.

Much work still remains in the areas of development and utilization of aqueous tape casting technology. While some manufacturers, both large and small, have already made the change to water, many others have not found it to be a worthwhile or even possible option. Tape thickness, processing speed, tape performance, and ease of manufacture are all issues that weigh upon the final decision of which solvent to choose. We have no doubt that further research will be done, further pressures will be applied, and many more discoveries will be made before organic solvent–based tape casting is phased out. Perhaps it never will be phased out entirely.

APPENDIX 1

Model Formulations and Procedures

As a part of this book we have been advised by almost everyone we approached to add a section on model formulations that researchers can use to begin looking at tape casting for their own applications. Although we have cast most ceramic materials and several metallic compositions from A to Z (alumina to zirconia), there are only a few that we are permitted to release that are in the published literature. We will share those formulations and procedures for preparation of tape casting slips in this appendix. We have tried to include oxides, nitrides, carbides, and metallic batch formulations using binder systems that can be sintered in both oxidizing and reducing atmospheres. Slight modifications in solvent content may have to be made to attain the proper viscosity for the thickness being cast. A separate section for aqueous-based tape casting formulations is also included.

A.1 SOLVENT-BASED FORMULATIONS

Oxides

Low Surface Area, Oxidizing Atmosphere Sintering:

The first category we will cover will be oxides that are sintered in an oxidizing atmosphere and that have a relatively low surface area. The model formulation selected is a polyvinyl butyral-based binder system using a 94 wt% aluminum oxide powder with an alumina surface area of 3.3 m^2/g.

Batch Formulation

Component	Composition (wt%)	Function
Part 1:		
Aluminum oxide	57.39	
Clay (kaolin)	1.54	
Talc	2.78	
Menhaden fish oil, blown Z-3	2.47	Dispersant
Xylenes	14.51	Solvent
Ethyl alcohol, 95% denatured	14.51	Solvent
Part 2:		
Butylbenzyl phthalate, S-160	3.09	Plasticizer
Poly(vinyl butyral), grade B-98	3.70	Binder

Batching Procedure

The procedure must be followed exactly to yield the best possible tape-cast product. This procedure is as follows:

1. Dry the alumina powder at 90 to 100°C for at least 24 hours.
2. Weigh the fish oil and dissolve in the xylenes and add to the ball mill (size depends upon the batch size). The mill should be one-third filled with grinding media of the appropriate size for the mill. The media and the mill composition should be selected based upon the material being processed.
3. Weigh and add the ethyl alcohol.
4. Weigh and add the inorganic powders to the mill.
5. Dispersion mill by rolling at approximately 58 rpm for 24 hours.
6. Weigh and add the plasticizer.
7. Weigh and add the binder.
8. Mix to dissolve and homogenize for an additional 24 hours.
9. Pour and de-air the slip at 635 mm of mercury vacuum for 8 minutes (or longer if the batch size is greater than 10 l).
10. Check the viscosity using a Brookfield viscometer, RV-4 spindle at 20 rpm for quality control.

At this point the slip is ready to cast. Casting is done on a silicone-coated Mylar carrier in a continuous tape casting machine. The doctor blade gap setting, speed of casting, etc. are determined by the tape thickness desired as an end product.

High Surface Area, Oxidizing Atmosphere Sintering

This category will be represented by a 99.5 wt% aluminum oxide powder with a surface area of 11.6 m^2/g. It also uses a poly vinyl butyral binder system.

Batch Formulation

Component	Composition (wt%)	Function
Part 1:		
Aluminum oxide	61.94	
Menhaden fish oil, blown Z-3	1.24	Dispersant
Xylenes	15.31	Solvent
Ethyl alcohol, anhydrous denatured	15.31	Solvent
Part 2:		
Poly(vinyl butyral), grade B-98	3.09	Binder
Butylbenzyl phthalate, S-160	1.55	Plasticizer (Type I)
Poly(alkylene glycol)	1.55	Plasticizer (Type II)

Batching Procedure

The procedure must be followed exactly in order to yield the best possible results in the tape-cast product.

1. Dry the alumina powder at 90 to 100°C for at least 24 hours.

2. Weigh and dissolve the fish oil in the xylenes and add to a mill jar that will accommodate the alumina batch desired. The mill should be one-third filled with grinding media of the appropriate size for the mill: for example, 1.27 cm cylindrical media for a one-liter mill. As mentioned previously, the media and mill composition should be selected based upon the material being processed; an alumina mill with alumina media would be used for this formulation.

3. Weigh and add the ethyl alcohol.

4. Weigh and add the hot alumina powder.
5. Dispersion mill by rolling at 58 rpm for 24 hours.
6. Weigh and add the plasticizers.
7. Weigh and add the binder.
8. Mix and homogenize by rolling for an additional 24 hours at 58 rpm.
9. Pour and de-air in a vacuum chamber with agitation at 635 mm of mercury for a minimum of 8 minutes (time depends upon the batch size).
10. An option at this point would be to filter the slip to remove agglomerates that may be left. Usually 50 to 100 μm filtration is sufficient.
11. Check the viscosity using a Brookfield viscometer, RV-4 spindle at 20 rpm for quality control.

At this point the slip is ready for tape casting. The carrier of choice is silicone-coated Mylar in a continuous tape casting machine. The blade gap, casting speed, etc. are selected to yield the tape thickness desired.

Wide Range of Surface Areas, Nonoxidizing Atmosphere Sintering

This formulation works well for materials with surface areas in the 3 to 10 m^2/g range that will be sintered in a nonoxidizing atmosphere such as nitrogen, hydrogen, or argon. It can be used in an oxidizing atmosphere as well, but it includes a binder that "unzips" or evaporates at elevated temperatures and does not need oxygen for its removal. The material selected for this formulation is an alumina powder with a surface area of 8.2 m^2/g and a D_{50} of 0.5 μm.

Batch Formulation

Component	Composition (wt%)	Function
Part 1:		
Methyl ethyl ketone	7.26	Solvent
Ethyl alcohol, 95% denatured	7.26	Solvent
Polyester/polyamine copolymer	1.04	Dispersant
Part 2:		
Alumina	69.40	
Part 3:		
Butylbenzyl phthalate, S-160	3.47	Plasticizer
Acrylic copolymer: methyl ethyl ketone (50:50 wt%)	11.57	Binder

Note: The acrylic is an ethyl methacrylate, Grade B-72

Batching Procedure

The batching order and procedure must be followed exactly in order to produce the results desired in both the slip and the final tape.

1. Weigh and add the dispersant, the MEK, and the ethyl alcohol to a mill jar that is one-third filled with grinding media of the appropriate composition and size for the material being processed.
2. Mill for 12 to 24 hours until the dispersant, which is in solid form, is dissolved.
3. Weigh and add the alumina, which has been dried at 90 to 100°C for at least 24 hours.
4. Dispersion mill by rolling at 60 rpm for 24 hours.
5. Weigh and add the plasticizer and the binder solution.
6. Roll for an additional 24 hours at 60 rpm to mix and homogenize.
7. Pour and de-air in a vacuum chamber at 635 mm of mercury for 8 (or more, depending on the batch size) minutes.
8. Check the viscosity using a Brookfield viscometer, RV-4 spindle at 20 rpm.

The slip is then ready to cast on a silicone-coated Mylar carrier in a continuous tape casting machine. The doctor blade gap, casting speed, etc. are adjusted to yield the desired tape thickness. This formulation has been used to produce the highest green bulk density achieved for an alumina tape: 2.81 g/cc.

Carbides

High Surface Area, Reducing or Neutral Atmosphere Sintering

The formulation selected is a high-surface-area silicon carbide powder that can be sintered in a reducing atmosphere or a neutral atmosphere. The binder burnout has to be done in an oxidizing atmosphere below the oxidation point for the SiC. The powder had a surface area of > 15 m^2/g.

Batch Formulation

Component	Composition (wt%)	Function
Part 1:		
Silicon carbide	48.99	
Menhaden fish oil, blown Z-3	2.22	Dispersant
Xylenes	21.08	Solvent
Ethyl alcohol, anhydrous denatured	13.17	Solvent
Part 2:		
Poly(alkylene glycol)	5.23	Plasticizer (Type II)
Butylbenzyl phthalate, S-160	4.42	Plasticizer (Type I)
Poly(vinyl butyral), grade B-98	4.90	Binder

Batching Procedure

The following procedure must be followed exactly as specified to yield the proper slip and tape properties.

1. Weigh and dissolve the fish oil in the xylenes and add to a mill jar that is one-third filled with grinding media. The mill and media size and composition should be selected based upon the materials being processed and the degree of contamination that can be tolerated.

2. Weigh and add the ethyl alcohol and the SiC powder.

3. Dispersion mill for 24 hours at 60 rpm on a set of rollers.
4. Weigh and add the plasticizers and binder.
5. Roll at 60 rpm to dissolve the binder and homogenize the mix.
6. Pour and de-air in a vacuum chamber using agitation: 635 mm of mercury for 8 or more minutes (time depends upon the volume of slip).
7. Check viscosity using a Brookfield viscometer, RV-4 spindle at 20 rpm.

The slip is then ready for casting. The doctor blade gap setting and speed of casting will determine the final tape thickness. This cast is also made on a silicone-coated Mylar carrier.

Low Surface Area, Reducing Sintering Atmosphere

The material selected for this formulation was a relatively coarse tungsten carbide powder. The binder system was a standard acrylic polymer. Since the WC was very coarse (16 μm average particle size) and had such a high density (15.7 g/cc), the formulation was difficult to prepare. The tape casting had to be done immediately after the slurry was poured from the preparation jar.

Batch Formulation

Component	Composition (wt%)	Function
Tungsten carbide	91.07	
Dibutyl phthalate	0.55	Plasticizer
Acrylic copolymer: methyl ethyl ketone (50:50 wt% solution), ethyl methacrylate, grade B-72	7.29	Binder
Methyl ethyl ketone	1.09	Solvent, viscosity, modifier

Batching Procedure

It is not as critical to follow a specific order of adding the ingredients to the mixing container with batches such as this one, where all of the ingredients are added at the same time and the process is essentially a mixing procedure.

1. Weigh and add the binder solution, the plasticizer, and the additional solvent to a high-density polyethylene (HDPE) jar. NOTE: We use HDPE jars in many cases where one or two casts will be made and where there is little or no milling or dispersion milling taking place. The jars can be reused for the same material and can be discarded when the work is completed. For a heavy powder such as WC, the media will have to be as heavy as possible without introducing any contamination. In this case, the jar was one-third filled with zirconia media. Ideally, one should use tungsten or WC grinding media. The jar size and the size of the media should be selected based upon the size of the batch.

2. Weigh and add the tungsten carbide powder.

3. Mix by rolling at 60 rpm for 4 to 6 hours.

4. Pour and de-air in a vacuum chamber with agitation. The pressure is 635 mm of mercury and the time is 8 or more minutes, depending on the batch size.

5. Gently stir the mix after de-airing to redisperse any particles that may have settled out during de-airing.

The slip is now ready for casting. The doctor blade gap and speed of casting will be determined based upon the thickness of tape desired. The cast can be made on a silicone-coated Mylar carrier surface.

Nitride

Although we have worked with several nitrides, we have only selected one for publication here. This is a commonly used material in the electronics industry: aluminum nitride or AlN.

Relatively High Surface Area, Nitrogen Sintering Atmosphere

The AlN powder used in this formulation had a surface area in the 5 to 15 m^2/g range. The formulation was prepared with a water-free system, since the AlN is so reactive with water vapor.

Batch Formulation

COMPONENT	COMPOSITION (WT%)	FUNCTION
Part 1:		
Aluminum nitride	66.14	
Menhaden fish oil, blown Z-3	3.31	Dispersant
Xylenes	15.48	Solvent
Ethyl alcohol, anhydrous denatured	9.66	Solvent
Part 2:		
Poly(alkylene glycol)	1.72	Plasticizer, Type II
Butylbenzyl phthalate, S-160	1.72	Plasticizer, Type I
Poly(vinylbutyral), grade B-98	3.97	Binder

Batching Procedure

As with the other batches, the procedure should be followed exactly as written. The order of ingredient additions can be critical to the success of the casting batch.

1. Weigh and dissolve the fish oil in the xylenes and add to a mill jar that is one-half filled with zirconia grinding media. The size of the mill and media are determined by the batch size.
2. Weigh and add the anhydrous ethyl alcohol.
3. Weigh and add the AlN powder.
4. Dispersion mill by rolling at 60 rpm for 24 hours.
5. Weigh and add the plasticizers and the binder.
6. Mix to dissolve and homogenize by rolling at 60 rpm for 24 hours.
7. Pour and de-air in a vacuum chamber at 635 mm mercury for 8 or more minutes, depending on the volume of the batch. Agitate the slip during the de-airing process.
8. Measure the viscosity using a Brookfield viscometer, RV-4 spindle at 20 rpm.

At this point the slip is ready to cast. The thickness is determined by the doctor blade gap setting, speed of cast, etc. The formulation is meant to be cast on a silicone-coated Mylar carrier.

Metals

Relatively Coarse Particle Size, Reducing Atmosphere Sintering

The powder selected for this category is a coarse (-270 mesh) intermetallic compound: iron aluminide or FeAl. This powder is extremely difficult to process into a thin sheet; therefore, tape casting is an ideal way to produce the precursor sheet, which is then sintered and rolled to final density.

Batch Formulation

Component	Composition (wt%)	Function
Iron aluminide	88.13	
Methyl ethyl ketone	4.12	Solvent/viscosity modifier
Ethyl methacrylate co-polymer:MEK (50:50 wt. fraction)	7.05	Binder solution
Dibutyl phthalate	0.70	Plasticizer

Batching Procedure

This is another example of a system where there is little or no grinding of the powder and little or no dispersion necessary. The viscosity of the slurry determines the degree of settling of the metallic particles. All of the ingredients are therefore added to the mixing jar at the same time.

1. Weigh and add all of the ingredients to a high-density polyethylene jar that is one-fourth filled with cylindrical zirconia grinding media. The size of the jar and the size of the media are determined by the batch size being processed.

2. Mix to homogenize the slurry by rolling at 60 rpm for 24 hours.

3. Pour into a container and de-air in a vacuum chamber with agitation at 635 mm mercury for sufficient time for the volume being processed. For 4L or less, 8 minutes is sufficient.

4. Measure the viscosity using a Brookfield viscometer, RV-4 spindle at 20 rpm.

The slip is then ready to cast on a continuous moving silicone-coated Mylar carrier. The doctor blade gap and casting speed are set based upon the tape thickness desired.

Intermediate Particle Size, Reduction Sintering Atmosphere

The powder selected for this formulation is a pure spherical nickel powder that has a D_{50} particle size of 6.2 μm and a surface area of 0.31 m^2/g. The binder/plasticizer system is one that can be removed in either a slightly oxidizing atmosphere or in a nitrogen atmosphere.

Batch Formulation

Component	Composition (wt%)	Function
Part 1:		
Nickel powder	80.32	
Menhaden fish oil, blown Z-3	0.40	Dispersant
Xylenes	7.23	Solvent
Ethyl alcohol, 95% denatured	7.23	Solvent
Part 2:		
Butylbenzyl phthalate, S-160	2.41	Plasticizer
Poly(vinyl butyral), grade B-98	2.41	Binder

Batching Procedure

This is one case where a dispersion milling step is followed, even with a metallic powder. The powder was fine enough and the particle shape was such that rapid settling would take place if the dispersion characteristics were not optimized. This was done by using the dispersion test explained in Section 2.3.

1. Weigh and dissolve the fish oil in the xylenes and add to a high-density polyethylene (HDPE) mixing jar that is about one-quarter filled with alumina grinding media. The size of the jar and the media are determined by the batch size being processed.

2. Weigh and add the ethyl alcohol and the nickel powder.

3. Dispersion mill by rolling at 60 rpm for a minimum of 4 hours.

4. Weigh and add the plasticizer and the binder.

5. Roll at 60 rpm for 24 hours to dissolve the binder and to homogenize the mix.

6. Pour into a container and de-air in a vacuum chamber at 635 mm of mercury for 8 or more minutes, depending on the size of the batch.

7. Measure the viscosity using a Brookfield viscometer, RV-4 spindle at 20 rpm.

The slip is then ready to cast onto a silicone-coated Mylar carrier in a continuous tape casting machine. The doctor blade gap and speed of casting are determined by the ultimate thickness that has to be cast.

Relatively Coarse-Particle-Size Brazing Alloy, Very-Low-Temperature Binder Removal

This formulation was prepared for a nickel brazing alloy from which the binder/plasticizer has to be removed at temperatures below 350°C. The poly(propylene carbonate) binder system evaporates at temperatures below this limit.

Batch Formulation

COMPONENT	COMPOSITION (WT%)	FUNCTION
Part 1:		
Poly(propylene carbonate), grade Q-40	3.36	Binder
Propylene carbonate	0.50	Plasticizer
Methyl ethyl ketone	12.25	Solvent/viscosity modifier
Part 2:		
Braze alloy	83.89	

Batching Procedure

Once again we have used a different batching procedure. As with most metallic systems, there is no appreciable grinding or dispersion milling; therefore we have eliminated that step in the procedure.

1. Predissolve the Part 1 ingredients in a high-density polyethylene (HDPE) jar that is about one-quarter filled with zirconia grinding media. The size of the mixing jar and the size and composition of the grinding (mixing) media are determined by the material being mixed. Usually it is good practice to use grinding media that has a density that is higher than or very close to the material being processed, otherwise it will tend to float (and may actually float!) in the slurry.
2. Weigh and add the metal powder.
3. Mix by rolling at 60 rpm for about 4 hours.

4. Pour and cast. NOTE: *Do not de-air this formulation.* The plasticizer will evaporate at room temperature. Also, store the tape in a sealed bag to prevent the loss of plasticizer. This is good practice for any tapes that are going to be stored.

The slip is ready to be cast. Casting can be done on a moving silicone-coated Mylar carrier in a continuous casting machine. The doctor blade gap setting and speed of casting will determine the ultimate tape thickness produced.

A.2 AQUEOUS-BASED FORMULATIONS

There are basically two types of water-based formulations: (1) those prepared using emulsion binder systems and (2) those prepared with water-soluble binder systems. Examples of each are included in this section.

Acrylic Emulsions

High Surface Area, Oxidizing Atmosphere Sintering

The acrylic binder systems can be used in either oxidizing or nonoxidizing atmospheres during sintering. The binders are removed by "unzipping" or evaporation and do not require oxygen for removal. The representative system for this category is a 99.5 wt% aluminum oxide powder with a surface area of 8.2 m^2/g.

Batch Formulation (Taken from Reference 21 in Chapter 7)

Component	Composition (wt%)	Function
Part 1:		
Aluminum Oxide	68.91	
DI water	14.45	Solvent
Low-MW ammonium poly acrylate salt,grade D3005	0.60	Dispersant
Part 2:		
Acrylic emulsion, cross-linkable, grade B1050	6.88	Binder, 48% TS
Acrylic emulsion, low T_g, grade B1035	9.02	Binder, plasticizer, 55% TS
Polyoxyalkylenediamine, grade D400	0.14	Curing agent, 50% TS

NOTE: TS = total solids or active ingredients in wt%.

Batching Procedure

This procedure must be followed exactly to yield the best possible tape-cast product. The procedure is as follows:

1. Dry the alumina powder at 90 to 100°C for at least 24 hours.
2. Weigh the dispersant, D3005, and the water and add to a ball mill (size depends upon the batch size). The mill should be one-third filled with grinding media of the appropriate size for the mill. Usually a large master batch is prepared of the dispersed powder.
3. Weigh and add the hot alumina powder to the mill.
4. Dispersion mill at 80 to 100 rpm for at least 3 hours.
5. Pour the proper amount of dispersed slurry into a beaker.
6. Weigh the binder(s) and curing agent and slowly pour them into the beaker with slow stirring, using an overhead stirrer to ensure uniform mixing. *DO NOT CAVITATE OR ENTRAP AIR DURING THIS MIXING PROCEDURE.* Stir for 30 minutes.
7. Filter through a 100-mesh sieve.

At this point the slip is ready to cast. Casting is done on a silicone-coated Mylar carrier in a continuous tape casting machine. The doctor blade gap setting, speed of casting, etc. are determined by the tape thickness desired as an end product.

Low Surface Area, Oxidizing Atmosphere Sintering

The powder selected for this model formulation is an X7R capacitor material with a surface area of 3 m^2/g.

Batch Formulation

COMPONENT	COMPOSITION (WT%)	FUNCTION
Part 1:		
X7R powder	71.51	
Ammonium polyacrylate, grade D-3019	0.72	Dispersant
DI water	10.96	Solvent
Part 2:		
Acrylic emulsion, cross-linkable, grade B-1050	7.15	Binder, 48% TS

Polyoxyalkylenediamine, grade D-400	0.15	Curing agent, 50% TS
Surfactant, grade X-405	0.14	Surfactant, 70% TS
Acrylic emulsion, low T_g, grade B-1000	9.36	Binder/plasticizer, 55% TS

NOTE: TS = total solids or active ingredients (wt%)

Batching Procedure

1. The DI water and dispersant are added to a ball mill jar of the proper size for the batch to be processed. The milling media are also sized according to the batch. The pH is adjusted to 9.5 using a 28% ammonium hydroxide solution, then the powder is added.
2. Dispersion mill at 40 to 50 rpm for 24 hours.
3. In a separate container equipped with an overhead stirrer, add the B-1050, D-400, X-405, and the B-1000 *in sequence* using moderate stirring to form a binder mixture. Be sure to stir for a short period of time after adding each ingredient.
4. Pour the dispersed slurry into a container equipped with an overhead stirrer then add the binder mixture while stirring slowly. Use about a 30 rpm rate for 24 hours. *DO NOT CAVITATE THE MIXTURE OR INTRODUCE ANY AIR.*
5. Filter through a 30 μm sieve.

At this point the slip is ready to cast. Casting is done in a continuous tape casting machine on a silicone-coated Mylar carrier. The doctor blade gap setting, speed of casting, etc. are determined by the thickness of tape being cast.

Water-Soluble Binders

High Surface Area, Oxidizing Atmosphere Sintering

This category will be represented by a 99.5 wt% aluminum oxide powder with a surface area of greater than 10 m^2/g. The binder system is based upon methyl cellulose.

Batch Formulation (Taken from Reference 12 in Chapter 7)

Component	Composition (wt%)	Function
Part 1:		
Aluminum oxide	41.40	
Polypropylene glycol, grade P-1200	0.45	Plasticizer
Glycerin	1.35	Plasticizer
Ammonium polyelectrolyte, grade C	0.10	Dispersant
DI water	15.75	Solvent
Part 2:		
Methyl cellulose, grade 20-214	1.80	Binder
Hot DI water	18.00	Solvent
Cold DI water	21.15	Solvent

Batching Procedure

This procedure must be followed exactly as listed to yield the best possible tapes.

1. Weigh and add the ingredients of Part 1 to a ball mill of the proper size for the batch being prepared. The type and size of media are selected based upon the mill size. The media should fill about one-third to one-half of the volume of the mill.

2. Dispersion mill on rollers for 16 to 24 hours at a speed of about 60 rpm.

3. Disperse the methyl cellulose in rapidly stirring 90°C DI water and then add the room temperature water in Part 2.

4. Add this binder solution to the ball mill and mix for 1 to 2 hours.

5. Pour and filter through a 200-mesh sieve or pour into another container and slow roll for 24 hours (approximately 1 to 2 rpm) to remove the entrained air.

At this point the slip is ready to cast. Casting is done on a Mylar carrier in a continuous tape casting machine. The doctor blade gap setting, speed of casting, etc. are determined by the tape thickness desired as an end product. The wet:dry ratio for casting this slip is approximately 6.5:1.

High-Surface-Area Oxide Powder, Oxidizing Sintering Atmosphere

The powder used for this example is an iron oxide with a surface area of > 10 m^2/g. The binder system is a water-soluble polyvinyl alcohol.

Batch Formulation

COMPONENT	COMPOSITION (WT%)	FUNCTION
Part 1:		
Iron oxide powder	53.93	
DI water	22.35	Solvent
Part 2:		
Polyvinyl alcohol, grade 75-15	3.24	Binder
DI water	18.33	Solvent
Polyethylene glycol, grade 400	1.62	Plasticizer
Polypropylene glycol, grade P-1200	0.54	Defoamer

Batching Procedure

The procedure must be followed exactly to yield the best possible tape-cast product.

1. Weigh and add the Part 1 ingredients to a ball mill of the proper size for the batch being prepared. The media volume should be in the range of one-third to one-half of the mill volume. The size and shape of the media should be determined by the size of the mill.
2. Dispersion mill for 16 to 24 hours by rolling at about 50 to 70 rpm.
3. Prepare a 15 wt% polyvinyl alcohol solution by dissolving polyvinyl alcohol in DI water. Usually a larger batch is prepared so that the proper amount of solution can be weighed and added during the Part 2 preparation.
4. Weigh and add the polyethylene glycol, the polypropylene glycol, and the binder solution to the mill.
5. Mix for 16 to 24 hours by rolling at 50 to 70 rpm.
6. Pour and de-air in a vacuum desiccator at 635 mm Hg for 8 minutes.
7. Age without stirring for 3 hours before casting.

At this point the slip is ready to cast. Casting is done on a Mylar, D-grade carrier in a continuous tape casting machine. The doctor blade gap setting, speed of casting, etc. are determined by the tape thickness desired as an end product.

APPENDIX 2

Calculations for Theoretical Airflows

to Remain Under the Lower Explosive Limit (LEL) During Tape Casting in a Continuous Machine

The method of calculation was taken from the NFPA 86 Standard for Ovens and Furnaces, 1995 Edition, published by the National Fire Protection Association, 1 Batterymarch Park, PO Box 9101, Quincy, MA 02269-9101. More specifically, it was taken from Section 7.7.3 of that standard, which is entitled, "Method for Calculating Solvent Vapor Ventilation Rate."

The constants and variables used in the calculation are as follows:

1. One liter of water weighs 0.998 kg at 21°C.
2. Dry air at 21°C and 0.76 m Hg (mercury) weighs 1.199 kg/m^3.
3. *SpGr* = specific gravity of solvent (water = 1.0).
4. *VD* = vapor density of solvent vapor (air = 1.0).
5. LEL_T = Lower explosive limit expressed in percent by volume in air, corrected for temperature.
6. Ventilation factor of safety = four times volume calculated to be barely explosive.

In SI units this calculates as follows:

$$\frac{0.998 \times SpGr}{1.199 \times VD} = \text{m}^3 \text{ vapor/L of solvent} \qquad \text{(Eq. 1)}$$

$$\frac{\text{m}^3 \text{ vapor}}{\text{L solvent}} \times \frac{(100 - LEL_T)}{LEL_T} = \text{m}^3 \text{ of barely explosive mixture/L solvent.} \qquad \text{(Eq. 2)}$$

$$\frac{\text{m}^3 \text{ mixture}}{} \times 4 = \text{m}^3 \text{ of diluted mixture @ 25\% } \frac{LEL_T\text{/L of solvent}}{\text{evaporated in L solvent process}} \qquad \text{(Eq. 3)}$$

Substituting

$$\frac{4}{1.199} \times \frac{0.998 \times SpGr}{VD} \times \frac{(100 - LEL_T)}{LEL_T} = \text{m}^3 \text{ of safety ventilation air @ 25\% } LEL_T\text{/L of solvent evaporated} \quad \text{(Eq. 4)}$$

The volume of fresh air required for safety ventilation is obtained by multiplying the factor calculated in this last equation by the liters, or portions of liters, of solvent evaporated in the oven per minute. The final value will be expressed as:

m^3/minute of ventilation air per liter of solvent evaporated.

This value must then be corrected for the temperature of the exhaust stream exiting the oven with the result being expressed as:

standard m^3/minute ventilation air per liter of solvent evaporated within the oven.

SAMPLE CALCULATION

Assumptions:

1. 300 mm tape width.
2. 0.5 mm wet tape thickness (as cast thickness).
3. Casting speed = 500 mm/min.
4. Casting slip contains 60 volume % (23 wt%) methyl ethyl ketone.
5. Drying air temperature is 80°C.

Under these casting conditions, the volume of MEK introduced into the exhaust stream would be:

300 mm $\times$ 0.5 mm $\times$ 500 mm/min $\times$ 1 L/10^6 mm^3 $\times$ 0.60 (vol. fraction) = **0.045 L/min.**

Solvent properties for methyl ethyl ketone:

Vapor density = 2.42 (Air = 1)

Specific gravity = 0.80615 (water = 1)

Lower explosive limit (LEL_T) = 1.8%

Substituting these values into Equation 4:

$$4 \times \frac{0.998}{1.199} \times \frac{0.80615}{2.42} \times \frac{(100 - 1.8)}{1.8} = \underline{\mathbf{60.508\ m^3/L\ at\ 21°C\ @\ 25\%\ \mathit{LEL}}}$$

To correct for the 80°C air temperature, use the following calculation to obtain the multiplication factor:

$$\frac{\text{Actual air temperature} + 273°C}{21°C + 273°C} = \text{Correction Factor} \qquad \text{(Eq. 5)}$$

Therefore in the example:

$$\frac{80°C + 273°C}{21°C + 273°C} = \frac{353°C}{294°C} = 1.2$$

If access to the NFPA 86 Standard for Ovens and Furnaces, 1995 Edition is available this correction factor can be found in Table 7.5.1.

Applying the correction factor, we can calculate the corrected volume of fresh airflow necessary to maintain safe ventilation under the casting conditions specified. This is as follows:

$$60.508\ m^3/L \times 1.2 = \underline{\mathbf{72.61\ m^3/L\ at\ 80°C.}}$$

Then in order to calculate the volume of ventilation air per unit time, one must multiply this number by the actual volume of solvent (MEK) being generated. This is done as follows:

$$72.61\ m^3/L \times 0.045\ \text{L of MEK/ min} = \underline{\mathbf{3.267\ m^3/min}}$$

Therefore, in our hypothetical tape casting situation, an airflow rate of 3.267 m^3/min would be required as a minimum value for safe operation. This translates to 115.36 cubic feet/minute for those of us who still think in English units. Most machine manufacturers take these calculations into consideration when designing the minimum airflow for safety shutdown systems, and most machines we have observed have minimum airflows that are larger than this minimum value by a factor of 2 or more!

NOTE: Local and/or state fire safety codes may vary.

Glossary of Terms

We have tried to include in this glossary all terms used in this book that have obscure, confusing, or sometimes misleading definitions. Terms used commonly in both industry and academia that have multiple meanings or connotations have been included, along with a specific definition consistent for use throughout this book. Readers lacking degrees in ceramics or material science will also be able to find words included that ceramists often take for granted. We realize that these definitions are often used with different, sometimes multiple, meanings in other publications.

Absorbed: Incorporated into the body, or interior.

Acidic: Having or pertaining to pHs below 7. A higher concentration of H^+ than OH^-.

Acreage: Area. Specifically, the amount of usable area on the surface of substrate or package for mounting other components or circuitry.

Adsorbed: Chemically attached to the surface but not incorporated into the bulk, or body.

Agglomerate: *Soft agglomerate* or simply *agglomerate*: A group of particles that behaves as a single particle. Individual particles are held together by electrical charge, water at the particle contacts, or other short-range forces. *Hard agglomerate* or *aggregate*: A group of particles joined into a single unit by partial sintering during calcination or some other strong bonding mechanism.

Alkali / Alkaline: Having or pertaining to pHs greater than 7. A higher concentration of OH^- than H^+. See also Basic.

Alumina: Al_2O_3. A commonly used oxide of aluminum.

Anhydrous: Literally, "without water." Often used to describe systems with negligible water content.

Anionic: Adjective describing a functional ion with a negative charge.

Aqueous: Literally, "in water."

Ash: Residue after burning. Ash left by organic polymers or other additives often contains trace amounts of intermetallic ions, left-over carbon and various other atomic species.

Azeotrope: A blend of liquids, fully soluble in each other, which behaves as a single liquid. An azeotropic mixture has a single boiling point, and a single vapor pressure, and it retains the original component ratio during evaporation regardless of the individual component evaporation rates.

Backscatter: (Backscattered image or electrons) An occurrence and/or analysis technique used in scanning electron microscopy (SEM).

Ball mill: Equipment consisting of two parallel rollers, often covered with rubber or other soft material, which rotates a mill jar. Rotation of the mill jar thoroughly mixes (or mills) the contents within.

Basic: Having or pertaining to a pH greater than 7.

Batching: Adding formulation components or "recipe ingredients" to the ball mill jar. "Batching a mill." Also called "charging."

Bernoulli principle: Named after the man who discovered this phenomenon, this is the principle that describes the lowering of pressure in a tube or channel when a gas (often air) moves quickly across the opening at a 90° or greater angle. For a simple check, put a straw in a glass of water, blow sideways across the dry end, and watch the water rise part way up the straw.

Binary system: A system containing two parts.

Bisque: Partially sintered, porous. Bisque ceramics are pieces that have not fully densified to the lowest volume.

Blade gap: Distance between the doctor blade bottom and the top of the casting surface. Also called blade height.

Blank: *verb*—Cut, stamp, or punch a part out of the cast tape.
noun—The part that was cut out.

Blown oil: Oil that has been oxidized. The oxidizing process sometimes involves blowing air or oxygen bubbles through the oil, hence the name.

Burn off: To oxidize, evaporate, or otherwise decompose one or more components by heating. This term is used whether or not the components actually burn or volatilize in any other manner.

Butvar®: Poly(vinyl butyral).

Calcining: This term literally means heating. It typically implies a heat treatment for reasons other than densifying the material. Calcining is performed to promote diffusion and the resulting homogenization of high-temperature materials, to burn off unwanted organic components, or to react two or more materials to form a single material.

Camber: Curvature or warpage. Substrates and laminates are often domed or cupped after firing. The resulting curvature or warp is called camber.

Carcinogen: Cancer-causing agent.

Card(s): See Blank

Carrier: Also *Carrier film* or *Web*. The surface upon which slurry is deposited and held through the drying step.

Case hardening: Drying of the outer layer of a suspension in such a way as to block the evaporation of liquid from the interior. The dry layer acts as a hard crust or case, which hinders drying.

Cationic: Adjective describing a functional ion with a positive charge.

Cavitation: Introduction of shear or tension into a liquid high enough to cause evaporation of the liquid. Cavitation forms bubbles within the liquid.

cc: Common abbreviation for cubic centimeter(s).

Charging: 1. Imparting a net electronic charge, positive or negative.
2. Adding components to a mill jar: "charging a mill." See also *Batching.*

Cold trap: A device that lowers the temperature of a gas, forcing the condensation of solvent (water or other) vapor. This can be used to reclaim solvents from the tape casting machine's exhaust.

Continuous phase: A phase or compound arranged in such a way as to have all sections connected. A material arranged with uninterrupted connections from one area to any other.

cP: Abbreviation for centipoise or 1/100 Poise. A unit of measure for viscosity.

Cracked ammonia: Also known as dissociated ammonia or DA. A reducing atmosphere gas that is produced from NH_3 by electrical dissociation into 75% H_2 and 25% N_2.

Cross-link: Attachment of one polymer chain to others, creating a three-dimensional, chemically bonded network of polymer.

Crow's foot crack: A nucleus from which three cracks propagate. This type of crack typically looks like a bird's footprint, hence the name.

D_{10}: Threshold diameter dimension, calculated or measured, at which 10 wt% of the powder particles are "finer than." Ten weight percent of the powder will fall through a hole of diameter D_{10}.

D_{50}: Threshold diameter dimension at which 50 wt% of the powder particles are "finer than."

D_{90}: Threshold diameter dimension, at which 90 wt% of the powder particles are "finer than."

De-air: Remove air from. To remove air from a prepared suspension or slip.

Deflocculant: An additive, often an oil or short-chain polymer, that separates or keeps separate the particles in a suspension. In some cases, with some materials, acids or bases can be used to deflocculate a slurry.

Densification: Reduction of volume of a given mass of material, thereby increasing the material's mass per unit volume. Done after removal of organic additives, this step is typically performed by heating to high temperatures. While this term would also apply to the technique of green tape calendering, ceramists tend to reserve the use of this word to the final firing process. See also *Sintering*.

Density: The mass of an object divided by its volume. Variations are created by defining the volume measuring technique. These different techniques of measuring or calculating volume give rise to: green bulk density, fired bulk density, apparent density, true density, envelope density and theoretical density.

Desiccator: A sealed container. Desiccators are used to create desired local environments such as partial vacuum or low humidity. Since containers to hold partial-vacuum environments require special design considerations, these containers are commonly referred to as "vacuum desiccators."

Dielectric breakdown: The conversion of an insulating dielectric material to an electrically conductive material, typically as a direct result of high applied electric field.

Dielectric constant: Also called relative permittivity or *K*. Represented by the variables ε_R and κ. A measure of a material's ability to store charge under the influence of an applied electric field. Values are given in comparison to a perfect vacuum or "free space," which has a dielectric constant of 1 by definition.

Dielectric strength: The maximum electric field a material can experience before undergoing dielectric breakdown.

Dilatant: Rheological state in which a fluid's viscosity increases with increasing shear stress.

Dislocation pinning: Stiffening mechanism (typically experienced in metallic systems) in which defects hinder the motion of dislocation planes. For a more in-depth discussion, see a metallurgy text.

Dispersant: Additive that retards settling of particles in suspension. The ideal dispersant would not only retard but prohibit settling. Dispersants function in two main ways, steric hindrance and ionic repulsion.

Dispersion: A stable (nonsettling) suspension of particles in a fluid medium. This term also generally implies not only suspension of particles but also deflocculation of the particles.

Dispersion mill: Jar mill charged with solvent, dispersant/deflocculant, and powder. This milling step is done for the sole purpose of creating a dispersion of the powder in the solvent vehicle. The term implies that binder and plasticizer have not yet been added to the slip.

DLVO theory: Theory explaining the balance of attractive and repulsive forces between particles in suspension. This theory is often used to explain the stability of a deflocculated suspension or the lack thereof. Acronymed for the four men who discovered and modeled this interaction: Derjaguin, Landau, Verwey, and Overbeek.

Doctor blade: Gating device used to define wet-tape cast height. Also used to refer to the gate itself.

Dopant: Small amount of a secondary material added to the primary material in order to create or enhance desired properties or characteristics. Typically called a *dopant* if addition is intentional and an *impurity* if it is unintentional.

Downstream: Chronologically later in the process.

Dry thickness: Average thickness of the tape-cast layer after the solvent has been evaporated.

Emulsion: Particles or droplets of one material stably suspended (we hope) in a fluid in which they are not soluble.

EPA: Environmental Protection Agency. U.S. governmental agency that uses various techniques to encourage individuals or corporations to refrain from polluting the environment.

Equivalent spherical diameter: Diameter of a perfect sphere having the same surface area as the powder being analyzed.

FBD: Fired bulk density. Mass of the object after firing divided by its geometric volume after firing.

Ferrite: Material based upon a form of iron oxide (FeO, Fe_2O_3, or Fe_3O_4). The term typically implies magnetic behavior or response. The term can also be used to describe materials that show magnetic responses similar to those of known iron-containing compounds.

Film: Thin layer.

Firing: Heating in a furnace or kiln. See *Sintering*.

Flexibility: Ability to bend under applied stress without permanently deforming (breaking is considered a permanent deformation). This term also implies the ability to return to the original state upon application of the opposite stress.

Floc: A group of particles loosely bound together.

Flocculate: To cause flocs to form. Flocculation can be caused by the addition of certain ionic species (depending on the system), temperature changes, or the application of electric or magnetic fields, among other influences. An additive used to create flocs is called a *flocculant*.

Fluid nature: See *Rheology*.

Free handling: Handling of a tape cast layer, whether manually or automatically, after the tape has been removed from the carrier.

Free water: Water in a tape casting formulation that starts the batching process as pure water. Water used to dissolve the dispersant and disperse the powder is free water. Water added as part of an emulsion or binder solution is not free water.

Gap: See *Blade gap*.

GBD: Green bulk density. Mass of the object before firing divided by its geometric volume before firing.

GOOD density: Green oxide-only density. Mass of the oxide (or other) particles in an object before firing divided by its geometric volume before firing. Can be calculated by multiplying GBD by the weight percentage of nonvolatiles in the object.

Grain growth: Increase in grain size during densification. Large grains grow at the expense of smaller grains.

Grandma's Cooking Pot: An easily remembered story to remind us not to place the limitations or freedoms of one circumstance on another circumstance where they may no longer be valid. See Section 4.2.

Green: Formed but not yet fired.

Green thickness: Thickness of the green tape-cast layer. See also *Dry thickness*.

Hydrate: To incorporate water. Water can be incorporated by absorption, adsorption, or simply by wetting surfaces and contact points.

Hydrophobic: Not liking water. Hydrophobic chemicals tend to ball up on themselves or coat surfaces of water in an attempt to get away from the water. Powders are occasionally referred to as hydrophobic if the presence of water or water vapor can damage them. Aluminum nitride, for example, may be called hydrophobic, since it reacts with water to form ammonia and alumina.

Hydroxyl/Hydroxylate: Contraction of Hydrogen-Oxygen. Made of or containing OH^- groups. A single $-OH^-$ group. These are often found on particle surfaces due to a reaction with atmospheric humidity or other water source.

Hygroscopic: Moisture loving; readily absorbing water (including atmospheric humidity).

IEP: Isoelectric point. The pH at which no net surface charge exists on particles in suspension. Sometimes referred to as point of zero charge or PZC.

Immersion density: The density of an object as measured using the Archimedes principle of fluid displacement. This is normally considered a most accurate gauge of fired density, since it accounts for surface roughness and non-ideal geometries. This technique can be used to calculate envelope density, apparent density, percent open porosity, and percent closed porosity.

In-line: As part of another processing step. Incorporation of one processing step into the flow of another step, e.g., inspection done at the exit of the drying chamber while the tape is still in motion.

Interstitial: "In between." A small particle can sit in the cavity created by the packing of larger particles. The cavity or site where the small particle sits is considered an interstitial vacancy.

Ionic repulsion: Used in reference to a dispersion or deflocculation mechanism, this term refers to the creation of like charges on particle surfaces, resulting in electrostatic repulsive forces. These forces then keep the particles apart and in suspension.

Jar mill: Synonym for ball mill. See *Ball mill*.

Laminate: *verb*—To adhere multiple layers of tape together to form, for all intents and purposes, a single part.
noun—The resulting part after lamination.

Leather-hard state: The state of a body, typically clay or clay-like, in which it resembles leather in flexibility, stiffness, and ductility.

Media: Small pieces of material used inside a mill jar to increase mixing energy and efficiency. Commonly used materials include alumina, stabilized zirconia, flint pebbles, steel, chrome-plated steel, boron carbide, silicon carbide, and plastic-coated versions of these. Common shapes include spherical, cylindrical, radius-end cylinders, and long rods. Sizes range from 1 mm or less to over 5 cm.

MEK: Acronym for methyl ethyl ketone.

Memory: Parts that are deformed plastically will often retain localized stress gradients in the area around the deformation. This stored stress can remain even after the part is returned to its original form. The local stresses can manifest themselves during firing by returning the part to the deformed shape. This phenomenon is personified to the extent that the part is considered to "remember" where and how it was deformed. Reverting to a previously experienced state is thus called memory. One common example of memory can be seen in the evolution of fingertip-sized indents in thick substrates. Rough handling of low–yield stress tapes can result in memory defects.

Menhaden fish: *Brevoortia tyrannus* or *munnawhatteaug*. An inedible fish found in American Atlantic and Gulf waters. Used for fertilizer, oil, or bait.

Metaphor: A figure of speech used for comparison. The ascribing of a characteristic or trait of one object to another object to describe similarity.

Mica: One of a naturally occurring group of minerals. An aluminosilicate-based mineral known for well-defined, weak cleavage planes that facilitate separation into thin layers with relatively large area.

Micron: Abbreviation for micrometer or one-millionth of a meter.

Mil: A strange mix of metric and English units standing for "milli-inch." 1 mil = 0.001 inch.

Mill (Jar mill, Mill jar): Short for Jar mill. Driven rollers that are used for rotating a mixing vessel or "mill jar." Mills (rollers) can be found in many sizes and configurations, single or multiple tier, long enough for only one jar or multiple jars. The term *mill* is often used loosely to refer to the jar itself. Very high-capacity jars are often built with their own stands and drive mechanisms in a single unit.

Miscible: Mixable. Describes full mixing of two fluids without dissolution of one into the other. Soluble = dissolves in, miscible = mixes into. The converse, immiscible, refers to a lack of mixing ability, as seen in the case of oil and water.

Monolayer: One layer. Often used loosely to describe an extremely thin layer, since a single layer is hard to measure or verify.

Mosaic tile model: A metaphorical tool to help visualize a green cast tape. Mosaic tiles and some other small tiles sold for bathrooms, kitchens, etc., are sold in sheets. Each tile in these sheets is bound to the others by plastic or metal wire. Seeing the tape-cast sheet as ceramic particles held together by binder wires is useful in understanding and consequently designing the cast tape to suit individual applications. This is covered in depth in Chapter 2.

Mother Nature: A fictional entity created or imagined to serve as a personified reason that things happen for reasons we don't want to admit we don't know. Also serves as a politically correct catch-all that doesn't offend people's religious beliefs.

MPa: Mega Pascal. One million pascals. A unit of pressure.

mPa·s: Milipascal-second. A unit of viscosity.

Non-aqueous: Not based on water.

OSHA: Occupational Safety and Health Agency. Federal agency that sets minimum guidelines for worker safety. Guidelines set by OSHA are legally binding and enforceable in court.

Oxidize: To chemically react with oxygen. Aluminum (Al) oxidizes to become alumina (Al_2O_3). Iron (Fe) oxidizes to become rust (Fe_2O_3, FeO, or Fe_3O_4).

Particle size distribution (PSD): A way of describing primary particle sizes in a batch of powder. The number of particles in a given size range, or the weight of particles that fall in that range, is measured and reported. A narrow distribution has a fairly uniform particle size with little variation between one primary particle and another. A wide distribution has a wide range of particle sizes. Other distribution types occur or are created, such as bimodal or trimodal. Changes in PSD can affect packing during drying (green density), cause or prevent density gradients in the green tape, and affect sintering extent and uniformity.

Pebble mill: Mill jar charged with pebbles, usually flint, in place of other milling media choices.

pH: Quantitative measurement of the balance of H^+ and OH^- ions in a fluid.

Phase separation: The physical separation of two insoluble, immiscible materials. Pour oil into a pail of water and you will see a perfect example of phase separation.

Pinhole: A tiny hole in the tape, typically the result of air bubbles bursting during drying.

Plasticity: The ability to deform plastically. Ability to irreversibly deform under an applied stress by motion within the tape matrix. Plastic deformation causes rearrangement within the tape structure, and thus the tape does not revert to its original shape when the applied stress is removed.

Precursor (ceramic): One part of the desired final body. An additive that reacts with other precursors at a later time (often during firing) to form the desired chemistry.

Primary particle: One individual particle. The smallest volumetric pieces a ceramic powder can be separated into without requiring cleavage of the crystal. In comparison, a floc is a loosely bound group of primary particles, and an aggregate is a partially fused group of primary particles.

Pseudoplastic: Shear thinning. This term is used to describe a slip that becomes more fluid when shear stress is applied. The shear thinning behavior is totally reversible in a pseudoplastic slip. Mustard and catsup are commonly seen pseudoplastic fluids.

psi: Acronym for pounds per square inch. A unit of pressure.

PVA: Acronym for polyvinyl alcohol.

Pyrolysis: A reaction or other change in chemical makeup due to the application of heat.

Reduce/Reduction The opposite of oxidize or oxidation. To remove oxygen.

Resistivity: A material property quantitatively reflecting the material's ability to retard electron motion. The component of electrical resistance that is due solely to the material and not its size or shape. The mathematical inverse of conductivity.

Rheology: The study of material flow. Used in the ceramic industry to describe the reaction of a fluid to an applied stress. While some slips may increase in viscosity upon stirring (dilatent rheology), others will decrease in viscosity (pseudoplastic rheology). The third major rheological category, Newtonian, describes a fluid whose viscosity does not change with applied shear stress.

Rough pump: A vacuum pump designed to pull a low-level vacuum (25 to 28 in. Hg) quickly.

SEM.: Acronym for scanning electron microscope.

Shear: Offset antiparallel forces. This term describes a type of applied force. Perhaps the easiest way to understand the definition is with a comparison. Tensile force describes pulling apart like stretching a rubber band. Compressive force describes pushing together like squeezing a rubber ball. Shear force describes sliding along like windshield wipers or rubbing your hands together. On a horizontal surface, compression is pushing down, tension is pulling up, and shear is pushing or pulling horizontally.

Shot: A specific type of milling or grinding media. Metal balls are often used instead of ceramic media when processing iron oxide–containing powders. Since any media or mill jar wear becomes part of the powder formulation, it is a good idea to use media and jar mills that are made of substances already in your powder formulation. Ferrite batches are often milled in a steel jar (steel mill) with steel balls (steel shot). This term probably stems from the physical similarity to shot (as in buckshot, shotgun, etc.).

Silane: Any member of a family of silicone-containing copolymers. These copolymers were designed to aid in adhesion between dissimilar materials. Some researchers have also used silanes as dispersion and wetting agents. These materials are often found to be cost prohibitive.

Simile: A comparison, typically, using the words like or as.

Sintering: Densification or consolidation by heating without melting.

Skin: (As it pertains to tape casting) A solid or very high-viscosity layer on the exposed surface(s) of a slurry due to evaporation of the solvent vehicle. A similar phenomenon may be observed by leaving vanilla pudding in the refrigerator uncovered overnight.

Slip or slurry: Suspension of powder in a liquid medium. These terms are often used interchangeably, but some would argue that *slip* refers specifically to clay suspensions.

Solids loading: The amount of solid material suspended in or 'loaded into' a slurry.

Specific gravity: Identical to density but used when referring to liquids. The weight of a certain volume of liquid, slip, slurry, or suspension. Water, by definition, has a specific gravity of 1 gram per cubic centimeter or 1 gram per milliliter at a temperature of 4°C and one atmosphere pressure.

Spindle: (as used in this book) Attachment used with a rotational viscometer. Spindles are available in various shapes and sizes for accurate measurement over a wide range of viscosities.

Steric hindrance: Physical separation of particles in a fluid suspension. Specifically, the hindrance to flocculation of a steric mechanism or arrangement of molecules. The steric hindrance mechanism typically relies on an adsorbed organic layer on particle surfaces that physically prohibits primary particles from touching each other.

Strain: A change in length.

Strength: The force holding an object together. Generally considered equal to the amount of stress required to break the object. In this instance, the stress required to significantly propagate a crack into the object. Many solid materials, and the majority of ceramics, display lower strength in tension than in compression. Strength is closely related to, but by no means the same as, "toughness."

Substrate: From *substratum*. The layer underneath. The base, whether ceramic, metal or plastic, upon which circuitry and additional components are placed.

Surfactant: Contraction of SURFace ACTive AgeNT. An additive used to change the chemical nature of a solid or liquid surface. Often used as a catch-all term to refer to an additive when its proper description does not readily come to mind or is not fully known or understood.

Synthetic: "Man-made". Synthesized by someone other than Mother Nature.

Tape-casting: Read the foreword of this book.

TEM: Acronym for transmission electron microscope.

T_g: Glass transition temperature. Crystalline solids have a sharp transition between solid and liquid phases called the melting point or melting temperature. Amorphous or noncrystalline solids such as glass do not have a firm melting point but instead soften until they are considered liquid. The glass transition temperature is used to describe the temperature range where glasses soften. Above T_g the material is considered liquid; below T_g it is normally considered solid.

TGA: Acronym for thermogravimetric analysis or thermogravimetric analyzer. Used for measuring changes in mass with changes in temperature.

Theoretical density: The highest possible density a material can have under the present conditions (temperature, pressure). A calculated value based upon perfect atomic arrangement.

Thixotrope: An additive that imparts thixotropy to a slip whether intentionally or unintentionally. This term is typically used only to refer to additions intended to induce thixotropy. Since the terms thixotropy and pseudoplasticity are so commonly interchanged, the term thixotrope is sometimes used, incorrectly, in reference to an additive that induces pseudoplasticity.

Thixotropic: A rheological behavior characterized by a reversible viscosity increase over time. Application of a high shear force will return the fluid to its "original" viscosity and restart the time-dependent behavior.

Toughness: Related to strength, toughness refers to the amount of energy required to break an object into two pieces. An object that begins to crack but does not fracture completely will still withstand an applied load, whereas an object displaying full or "catastrophic" failure cannot. Strengths being equal, a wooden board will be tougher than a bar of glass, since the wood will fail in stages (splintering along the grain) and the glass will break all at once.

Type I plasticizer: An additive used to soften the binder polymer. The Type I plasticizer is a solvent for the binder and as such works to lower T_g of the polymer. The only difference between the solvent(s) and the Type I plasticizer is seen in the speed of evaporation.

Type II plasticizer: An additive used to soften the green tape matrix. The Type II plasticizer is not thought to interact chemically with the binder polymer but acts mechanically to partially prevent formation of the 3-D polymer matrix. The Type II plasticizer is often oily as well, acting as a lubricant within the tape matrix.

van der Waals: (as in *van der Waals forces* or attraction) Named for their discoverer J. D. van der Waals (a Dutch chemist), these are relatively weak attractive forces between molecules.

Viscometer: Viscosity-meter. An instrument used to measure viscosity. For tape casting slips, a balance between accuracy, repeatability, cost, and ease of use is typically (but not always) achieved with a rotational viscometer.

WC: The chemical formula for tungsten (W) carbide (C).

Wet hydrogen: Hydrogen gas that has been partially saturated with water vapor. The introduction of water vapor is typically accomplished simply by bubbling the flowing gas through a water bath. The oxygen content (from water) is often high enough to oxidize carbon residue from organics but low enough to avoid unwanted oxidation of the ceramic or metallic parts.

Wet nitrogen: Nitrogen laden with water vapor. See Wet hydrogen.

Wet thickness: Thickness of the tape casting slurry after passing under the doctor blade. This thickness is related to but rarely ever equal to the blade gap. This term is often used interchangeably with Blade Gap anyway, since true wet thickness is difficult and/or expensive to measure.

x-**direction:** Along the casting direction. The direction of carrier motion.

y-**direction:** Across the width of the cast tape 90° to the x-direction but in plane with the tape.

Yield stress: Stress at which the slope of a stress vs. strain curve becomes nonlinear. This stress is typically characterized by the onset of irreversible strain. At the yield stress, deformation switches from purely elastic to a mixture of elastic and plastic. If a material (tape) is subject to stress above the yield stress, the tape dimensions will remain altered even upon removal of the stress.

***z*-direction:** Through the thickness of the tape. Normal to the tape plane.

Zeta potential: Related closely with DLVO theory. A measure of the electrical charge on the effective particle surface. A key parameter when relying on the ionic repulsion mechanism for dispersion/deflocculation.

Zipper bag theory: See description in Section 3.1.

References

FOREWORD

1. G. N. Howatt, R. G. Breckenridge, and J. M. Brownlow, "Fabrication of Thin Ceramic Sheets for Capacitors," *J. Am. Ceram. Soc.*, **30** [8] 237–242 (1947).

CHAPTER 1 - INTRODUCTION AND HISTORY

1. G. N. Howatt, R. G. Breckenridge, and J. M. Brownlow, "Fabrication of Thin Ceramic Sheets for Capacitors," *J. Am. Ceram. Soc.*, **30** [8] 237–242 (1947).

2. G. N. Howatt, "Method of Producing High-Dielectric High-Insulation Ceramic Plates," U.S. Patent 2,582,993, Jan. 22, 1952.

3. Model 1000-001 Dreicast Coating System, DreiTek, Inc., 1277 Linda Vista Drive, San Marcos, CA 92069.

4. J. L. Park, Jr., "Manufacture of Ceramics," U.S. Patent 2,966,719, Jan. 3, 1961.

CHAPTER 2—MATERIALS TECHNOLOGY AND SELECTION

2.1 Powders

1. R. D. Cadle, *Particle Size.* Reinhold Publishing Corp., New York, 1965.

2. T. Allen, *Particle Size Measurements,* 3rd ed. Chapman and Hall, London, 1981.

3. R. N. Katz, "Characterization of Ceramic Powders"; pp. 35–49 in Treatise on Materials Science and Technology, Vol. 9, *Ceramic Fabrication Processes.* Edited by F. F. Y. Wang. Academic Press, New York, 1976.

4. S. G. Malghan, "Characterization of Ceramic Powders"; pp. 63–74 in Engineered Materials Handbook, Vol. 4, *Ceramics and Glasses.* Edited by S. J. Schneider. ASM International, Newbury, OH, 1991.

5. T. Allen and R. Davies, "Modern Aspects of Particle Size Analysis"; pp. 721–46 in Advances in Ceramics, Vol. 21, *Ceramic Powder Science.* Edited by G. L Messing et al. The American Ceramic Society, Westerville, OH, 1987.

6. J. S. Reed, *Introduction to the Principles of Ceramic Processing*; pp. 185–251. John Wiley & Sons, New York, 1988.

7. D. J. Shanefield, *Organic Additives and Ceramic Processing*; pp. 91–130. Kluwer Academic Publishers, Boston, 1995.

8. D. J. Shanefield, *Organic Additives and Ceramic Processing*; pp. 115–26. Kluwer Academic Publishers, Boston, 1995.

9. D. J. Shanefield, *Organic Additives and Ceramic Processing*; pp. 109–10. Kluwer Academic Publishers, Boston, 1995.

10. B. C. Wood, "Particle Sizing"; pp. 83-89 in Engineered Materials Handbook, Vol. 4, *Ceramics and Glasses*. Edited by S.J. Schneider. ASM International, Newbury, OH, 1991.

11. R. E. Mistler, D. J. Shanefield, and R. B.Runk, "Tape Casting of Ceramics"; pp. 416–17 in *Ceramic Processing Before Firing*. Edited by G. Y. Onoda, Jr. and L. L. Hench. John Wiley & Sons, New York, 1978.

12. R. E. Mistler, D. J. Shanefield, and R. B. Runk, "Tape Casting of Ceramics," p. 419 in *Ceramic Processing Before Firing*. Edited by G. Y. Onoda, Jr. and L. L. Hench. John Wiley & Sons, New York, 1978.

13. D. J. Shanefield, *Organic Additives and Ceramic Processing*; p. 118. Kluwer Academic Publishers, Boston, 1995.

14. R. E. Mistler, "Grain Boundary Diffusion and Boundary Migration Kinetics in Aluminum Oxide, Sodium Chloride, and Silver," Sc.D. Thesis, Massachusetts Institute of Technology, Cambridge, MA, 1967.

15. S. Lowell, *Introduction to Powder Surface Area*. John Wiley & Sons, New York, 1979.

16. S. Lowell and J. E. Shields, *Powder Surface Area and Porosity*. Chapman and Hall, New York, 1984.

17. S. Brunauer, P. H. Emmett, and E. Teller, "Adsorption of Gases in Multimolecular Layers" *J. Am. Chem. Soc.*, **60,** 309 (1938).

18. E. M. Anderson, R. A. Marra, and R. E. Mistler, "Tape Casting Reactive Aluminas," *Am. Ceram. Soc. Bull.*, **76** [7] 45–50 (1997).

2.2 Solvents

19. P. Boch and T. Chartier, "Tape Casting and Properties of Mullite and Zirconia-Mullite Ceramics," *J. Am. Ceram. Soc.*, **74** [10] 2448–52 (1991).

20. R. E. Mistler, D. J. Shanefield, and R. B. Runk, "Tape Casting of Ceramics"; p. 417 in *Ceramic Processing Before Firing*, Edited by G. Y. Onoda, Jr. and L. L Hench. John Wiley & Sons, New York, 1978.

21. D. J. Cronin and K. A. McMarlin, "Tape Casting of Dielectric Substrates"; pp. 1963–72 in *Ceramics Today — Tomorrow's Ceramics*. Edited by P. Vincenzini. Elsevier Science Publishers B.V., Amsterdam, 1991.

22. H. Burrell, "Solubility Parameters," *Interchem. Rev.*, **14** [1] 3–16 (1955).

23. W. R. Cannon, J. R. Morris and K. R. Mikeska, "Dispersants for Nonaqueous Tape Casting"; pp. 161–74 in Advances in Ceramics, Vol. 19, *Multilayer Ceramic Devices*. Edited by J. B. Blum and W. R. Cannon. The American Ceramic Society, Westerville, OH, 1986.

24. K. R. Mikeska and W. R. Cannon, "Dispersants for Tape Casting Pure Barium Titanate"; pp. 164–83 in Advances in Ceramics, Vol. 9, *Forming of Ceramics*. Edited by J. A. Mangels and G. L. Messing. The American Ceramic Society, Westerville, OH, 1984.

25. C. Galassi, E. Roncarni, C. Capiani and P. Pinasco, "PZT-Based Suspensions for Tape Casting," *J. Eur. Ceram. Soc.*, **17**, 367–371 (1997). Elsevier Science Limited, 1996.

26. C. His, N. Chardon, R. Kuentzler, and S. Vilminot, "Elaboration and Characterization of $YBa_2Cu_3O_{7-x}$ Thick Tapes," *J. of Mat. Sci.*, **26**, 4829–35 (1991).

27. P. Boch and T. Chartier, "Understanding and Improvement of Ceramic Processes: The Example of Tape Casting," in *Ceramic Developments*, Mat. Sci. Forum, Vols. 34–36. Edited by C. C. Sorrell, B. Ben-Nissan. Trans Tech Publications Ltd., Switzerland, 1988.

28. W. R. Cannon, R. Becker and K. R. Mikeska, "Interactions Among Organic Additives Used for Tapecasting"; pp. 525–41 in Advances in Ceramics, Vol. 26, *Ceramic Substrates and Packages for Electronic Applications*. Edited by M. F. Yan et al. The American Ceramic Society, Westerville, OH, 1989.

29. J. C. Lin, T. S. Yeh, C. L. Cherng and C. M. Wang, "The Effects of Solvents and Binders on the Properties of Tape Casting Slurries and Green Tapes"; pp. 197–204 in Ceramic Transactions, Vol. 26, *Forming Science and Technology for Ceramics*. Edited by M. J. Cima. The American Ceramic Society, Westerville, OH, 1992.

30. R. J. MacKinnon and J. B. Blum, "Particle Size Distribution Effects on Tape Casting Barium Titanate"; pp. 150–57 in Advances in Ceramics, Vol. 9, *Forming of Ceramics*. Edited by J. A. Mangels and G. L. Messing. The American Ceramic Society, Westerville, OH, 1984.

31. J. A. Lewis, K. A. Blackman, A. L. Ogden, J. A. Payne and L. F. Francis, "Rheological Property and Stress Development During Drying of Tape-Cast Ceramic Layers," *J. Am. Ceram. Soc.*, **79** [12] 3225–34 (1996).

32. Y. Imanaka et al., Fujitsu Ltd., Kawasaki, Japan, "Process for Producing Multilayer Ceramic Circuit Board with Copper," U.S. Patent 4,679,320, Jul. 14, 1987.

33. J. L. Park, Jr., "Manufacture of Ceramics," U.S. Patent 2,966,719, Jan. 3, 1961.

34. J. S. Reed, *Principles of Ceramic Processing*. John Wiley & Sons, New York, 1988.

35. N. Kamehara, et al., Fujitsu Ltd., Kawasaki, Japan, "Method for Producing Multilayered Glass-Ceramic Structure with Copper-Based Conductors Therein," U.S. Patent 4,504,339 (Mar. 12, 1985).

36. *Azeotropic Data*, Advances in Chemistry Series, No. 6, American Chemical Society, Washington, D.C., 1952.

37. Eric West; private communication.

38. H. D. Kaiser, F. J. Pakulski and A. F. Schmeckenbecher, "A Fabrication Technique for Multilayer Ceramic Modules," *Solid State Technology*, May, 1972, pp. 35–40.

2.3 Surfactants

39. H. D. Kaiser, F. J. Pakulski and A.F. Schmeckenbecher, "A Fabrication Technique for Multilayer Ceramic Modules," *Solid State Technology*, May 1972, pp. 35–40.

40. R. G. Horn, "Particle Interactions in Suspensions"; pp. 58–101 in *Ceramic Processing*. Edited by R. A. Terpstra, P. P. A. C. Pex, and A. H. DeVries. Chapman & Hall, London,1995.

41. J. S. Reed, *Principles of Ceramic Processing*; Ch. 10. John Wiley & Sons, New York, 1988.

42. T. Hayashi, "Surface Chemistry of Ceramic Shaping Processes"; pp. 16–36 in *FC Annual Report for Overseas Readers*, Japan Fine Ceramic Association, Sept. 1991.

43. J. S. Reed, *Principles of Ceramic Processing*; p. 148. John Wiley & Sons, New York, 1988.

44. J. L. Park, Jr., "Manufacture of Ceramics," U.S. Patent 2,966,719, Jan. 3, 1961.

45. E. S. Tormey, "The Adsorption of Glyceryl Esters at the Alumina/Toluene Interface"; pp. 290–92. Ph.D. Thesis, Massachusetts Institute of Technology, Cambridge, MA, 1982.

46. E. S. Tormey, R. L. Pober, H. K. Bowen, and P. D. Calvert, "Tape Casting—Future Developments"; pp. 140–49 in Advances in Ceramics, Vol. 9, *Forming of Ceramics*. Edited by J. A. Mangels and G. L. Messing. The American Ceramic Society, Westerville, OH, 1984.

47. K. R. Mikeska and W. R. Cannon, "Dispersants for Tape Casting Pure Barium Titanate"; pp. 164–83 in Advances in Ceramics, Vol. 9, *Forming of Ceramics*. Edited by J. A. Mangels and G. L. Messing. The American Ceramic Society, Westerville, OH, 1984.

48. B. I. Lee and J. P. Rives, "Dispersion of Alumina Powders in Nonaqueous Media," *Colloids and Surfaces*, **56**, 25–43 (1991).

49. T. Hayashi, "Surface Chemistry of Ceramic Shaping Processes"; Fig. 1 in *FC Annual Report for Overseas Readers*, Japan Fine Ceramic Association. Sept. 1991.

50. R. J. MacKinnon and J. B. Blum, "Particle Size Distribution Effects on Tape Casting Barium Titanate"; pp. 150–57 in Advances in Ceramics, Vol. 9, *Forming of Ceramics*. Edited by J. A. Mangels and G. L. Messing. The American Ceramic Society, Westerville, OH, 1984.

51. R. E. Mistler, D. J. Shanefield, and R. B. Runk, "Tape Casting of Ceramics" p. 421 in *Ceramic Processing Before Firing*. Edited by G. Y. Onoda, Jr. and L. L Hench. John Wiley & Sons, New York, 1978.

52. D. J. Shanefield, *Organic Additives and Ceramic Processing*; pp. 241–43. Kluwer Academic Publishers, Boston, 1995.

53. Wally Meckes, Werner G. Smith, Inc.; personal communication

54. E. S. Tormey, "The Adsorption of Glyceryl Esters at the Alumina/Toluene Interface"; p. 289. Ph.D. Thesis, Massachusetts Institute of Technology, Cambridge, MA, 1982.

55. W. R. Cannon, J. R. Morris and K. R. Mikeska, "Dispersants for Nonaqueous Tape Casting"; pp. 161–62 in Advances in Ceramics, Vol. 19, *Multilayer Ceramic Devices*. Edited by J. B. Blum and W. R. Cannon. The American Ceramic Society, Westerville, OH, 1986.

56. K. R. Mikeska and W. R. Cannon, "Non-Aqueous Dispersion Properties of Pure Barium Titanate for Tape Casting," *Colloids and Surfaces*, **29**, 305–21 (1988).

57. M. Geho and H. Palmour III, "Sources of Sintering Inhibition in Tape-Cast Aluminas," *Ceramic Eng. & Sci. Proceedings*, **14** [11–12] 97–129 (1993).

58. R. N. Katz, "Characterization of Ceramic Powders"; p. 179 in Treatise on Materials Science and Technology, Vol. 9, *Ceramic Fabrication Processes*. Edited by F. F. Y. Wang. Academic Press, New York, 1976.

59. R. E. Mistler, "Tape Casting: The Basic Process for Meeting the Needs of the Electronics Industry," *Am. Ceram. Soc. Bull.*, **69** [6] 1022–26 (1990).

2.4 Binders

60. W. K. Shih, M. D. Sacks, G. W. Scheiffele, Y. N. Sun, and J. W. Williams, "Pyrolysis of Poly(Vinyl Butyral) Binders: I, Degradation Mechanisms"; pp. 549–58 in Ceramic Transactions, Vol. 1, *Ceramic Powder Science II*, A. Edited by G. L. Messing, E. R. Fuller, and H. Hausner, The American Ceramic Society, Westerville, OH, 1988.

61. G. W. Scheiffele and M. D. Sacks, "Pyrolysis of Poly(Vinyl Butyral) Binders: II, Effects of Processing Variables"; pp. 559–66 in Ceramic Transactions, Vol. 1, *Ceramic Powder Science II*, A. Edited by G. L. Messing, E. R. Fuller, and H. Hausner. The American Ceramic Society, Westerville, OH, 1988.

62. J. C. Williams, "Doctor Blade Process"; pp. 173–98 in Treatise on Materials Science and Technology, Vol. 9, *Ceramic Fabrication Processes*. Edited by F. F. Y. Wang. Academic Press, New York, 1976.

63. T. Chartier and A. Bruneau, "Aqueous Tape Casting of Alumina Substrates," *J. Eur. Ceram. Soc.*, **12**, 243–47 (1993).

64. E. A. Groat and T. J. Mroz, "Aqueous Processing of AlN Powders," *Ceramic Industry*, **140** [3] 34–38 (1993).

65. J. E. Schuetz, I. A. Khoury, and R. A. DiChiara, "Water-Based Binder for Tape Casting," *Ceramic Industry*, **129** [5] 42–44 (1987).

66. M. Hurley, Pac Polymers, Inc; private communication.

2.5 Plasticizers

67. G. O. Medowski, R. D. Sutch; personal communication, 1972.

68. T. Hayashi, "Surface Chemistry of Ceramic Shaping Processes"; pp. 16–36 in *FC Annual Report for Overseas Readers*, Sept. 1991, Japan Fine Ceramic Association.

69. W. R. Cannon, R. Becker, and K. R. Mikeska, "Interactions Among Organic Additives Used for Tapecasting"; pp. 525–41 in Advances in Ceramics, Vol. 26, *Ceramic Substrates and Packages for Electronic Applications*. Edited by M. F. Yan, et al. The American Ceramic Society, Westerville, OH, 1989.

70. E. H. Immergut and H. F. Mark, "Principles of Plasticization"; p. 6 in Advances in Chemistry Series, Vol. 48, *Plasticization and Plasticizer Processes*. Edited by R. F. Gould. American Chemical Society, Washington, DC, 1965.

71. T. Hayashi, "Surface Chemistry of Ceramic Shaping Processes"; Fig. 1 in *FC Annual Report for Overseas Readers*, Japan Fine Ceramic Association, Sept. 1991.

72. Santicizer® 160—Publication No. 2311534B, Monsanto Company. ®Registered Trademark of Monsanto Company.

73. R. A. Gardner and R. W. Nufer, "Properties of Multilayer Ceramic Green Sheets," *Solid State Technology*, (May 1974).

2.6 Organic Interactions

74. T. Hayashi, "Surface Chemistry of Ceramic Shaping Processes"; p. 17 in *FC Annual Report for Overseas Readers*, Japan Fine Ceramic Association, September 1991.

75. W. R. Cannon, J. R. Morris, and K. R. Mikeska, "Dispersants for Nonaqueous Tape Casting"; pp. 161–74 in Advances in Ceramics, Vol. 19, *Multilayer Ceramic Devices*. Edited by J. B. Blum and W. R. Cannon. The American Ceramic Society, Westerville, OH, 1987.

76. K. Blackman, "Processing of Non-Aqueous Tape-Cast Ceramic Layers," M.S. Thesis. University of Illinois, Urbana-Champaign, IL, 1996.

77. A. Roosen, F. Hessel, H. Fischer, and F. Aldinger, "Interaction of Polyvinyl Butyral With Alumina"; pp. 451–59 in Ceramic Transactions, Vol. 12, *Ceramic Powder Science III*, Edited by G. L. Messing, S. Hirano, and H. Hausner. The American Ceramic Society, Westerville, OH, 1990.

CHAPTER 3—MATERIALS PROCESSING: SLIP PREPARATION

3.1 Pre-Batching Powder Preparation

1. R. E. Mistler and D. J. Shanefield, "Washing Ceramic Powders to Remove Sodium Salts," *Am. Ceram. Soc. Bull.*, **57** [7] 689 (1978).

2. DeLaval Separator Co., Poughkeepsie, NY.

3. H. W. Stetson and W. J. Gyurk, "Alumina Substrates," U.S. Patent 3,698,923, Oct. 17, 1972.

4. J. P. Rives and B. I. Lee, "The Effect of Water on the Dispersion of Alumina in Nonaqueous Media"; pp. 45–57 in *Colloids and Surfaces, Vol. 56*. Elsevier Science Publishers B.V., Amsterdam, 1991.

5. M. D. Sacks and G. W. Scheiffele, "Polymer Adsorption and Particulate Dispersion in Nonaqueous Al_2O_3 Suspensions Containing Poly (vinyl butyral) Resins"; pp. 175–84 in Advances in Ceramics, Vol. 19, *Multilayer Ceramic Devices*. Edited by J. B. Blum and W. R. Cannon. The American Ceramic Society, Westerville, OH, 1986.

6. E. M. Anderson; private communication.

7. D. J. Shanefield, *Organic Additives and Ceramic Processing*; p. 139. Kluwer Academic Publishers, Boston, 1995.

8. Wally Meckes, Werner G. Smith, Inc.; personal communcation.

9. E. S. Tormey, "The Adsorption of Glyceryl Esters at the Alumina/Toluene Interface"; p. 289. Ph.D. Thesis, Massachusetts Institute of Technology, Cambridge, MA, 1982.

10. K. R. Mikeska and W. R. Cannon, "Dispersants for Tape Casting Pure Barium Titanate"; pp. 164–83 in Advances in Ceramics, Vol. 9, *Forming of Ceramics*. Edited by J. A. Mangels and G. L. Messing. The American Ceramic Society, Westerville, OH, 1984.

11. D. J. Shanefield, *Organic Additives and Ceramic Processing*; p. 248. Kluwer Academic Publishers, Boston, 1995.

12. K. Lindquist, et al., "Organic Silanes and Titanates as Processing Additives for Injection Molding," *J. Am. Ceram. Soc.*, **72** [1] 99–103 (1989).

13. E. A. Groat and T. J. Mroz, "Aqueous Processing of AlN Powders," *Ceramic Industry*, **140** [3] 34–38 (1993).

14. T. J. Mroz, "Aluminum Nitride," *Am. Ceram. Soc. Bull.*, **71** [5] 782–84 (1992).

15. D. J. Shanefield, *Organic Additives and Ceramic Processing*; p. 210. Kluwer Academic Publishers, Boston, 1995.

16. D. Hotza, et al., "Hydrophobic Aluminum Nitride Powder for Aqueous Tape Casting"; pp. 397–401 in Ceramic Transactions, Vol. 51, *Ceramic Processing Science and Technology*. Edited by H. Hausner. G. L. Messing, and S. Hirano. The American Ceramic Society, Westerville, OH, 1995.

3.2 Dispersion Milling

17. H. W. Stetson and W. J. Gyurk, "Alumina Substrates," U.S. Patent 3,698,923, Oct. 17, 1972.

18. *Handbook of Ball Mill and Pebble Mill Operation*; p. 7. Paul O. Abbe', Inc., Little Falls, NJ.

19. J. S. Reed, *Introduction to the Principles of Ceramic Processing*; p. 148. John Wiley & Sons, New York, 1988.

20. R. E. Mistler, D. J. Shanefield, and R. B. Runk, "Tape Casting of Ceramics"; p. 417 in *Ceramic Processing Before Firing*. Edited by G. Y. Onoda, Jr. and L. L. Hench. John Wiley & Sons, New York, 1978.

21. W. R. Cannon, J. R. Morris, and K. R. Mikeska, "Dispersants for Nonaqueous Tape Casting"; p. 171 in Advances in Ceramics, Vol. 19, *Multilayer Ceramic Devices*. Edited by J. B. Blum and W. R. Cannon. The American Ceramic Society, Westerville, OH, 1986.

3.3 Plasticizer and Binder Mixing

22. W. R. Cannon, J. R. Morris and K. R. Mikeska, "Dispersants for Nonaqueous Tape Casting"; pp. 161–74 in Advances in Ceramics, Vol. 19, *Multilayer Ceramic Devices*. Edited by J. B. Blum and W. R. Cannon. The American Ceramic Society, Westerville, OH, 1986.

23. K. R. Mikeska and W. R. Cannon, "Dispersants for Tape Casting Pure Barium Titanate"; pp. 164–83 in Advances in Ceramics, Vol. 9, *Forming of Ceramics*. Edited by J. A. Mangels and G. L. Messing. The American Ceramic Society, Westerville, OH, 1984.

24. J. E. Schuetz, I. A. Khoury, and R. A. DiChiara, "Water-Based Binder for Tape Casting," *Ceramic Industry*, **129** [5] 42–44 (1987).

25. T. Hayashi, "Surface Chemistry of Ceramic Shaping Processes"; pp. 16–36 in *FC Annual Report for Overseas Readers*, Japan Fine Ceramics Association, September 1991.

26. E. S. Tormey et al., "Tape Casting—Future Developments"; pp. 140–49 in Advances in Ceramics, Vol. 9, *Forming of Ceramics*. Edited by J. A. Mangels and G. L. Messing. The American Ceramic Society, Westerville, OH, 1984.

27. W. R. Cannon, R. Becker, and K. R. Mikeska, "Interactions Among Organic Additives Used for Tape Casting"; pp. 525–41 in Advances in Ceramics, Vol. 26, *Ceramic Substrates and Packages for Electronic Applications*. Edited by M. F. Yan et al. The American Ceramic Society, Westerville, OH, 1989.

28. D. J. Shanefield, "Competing Adsorptions in Tape Casting"; pp. 155–60 in Advances in Ceramics, Vol. 19, *Multilayer Ceramic Devices*. Edited by J. B. Blum and W. R. Cannon. The American Ceramic Society, Westerville, OH, 1986.

29. K. A. Blackman, "Processing of Nonaqueous Tape Cast Ceramic Layers," M.S. Thesis. University of Illinois, Urbana-Champaign, IL, 1996.

30. R. E. Mistler, D. J. Shanefield, and R. B. Runk, " Tape Casting of Ceramics"; p. 421 in *Ceramic Processing Before Firing*. Edited by G. Y. Onoda, Jr. and L. L. Hench. John Wiley and Sons, New York, 1978.

31. D. J. Shanefield and R. E. Mistler, "Fine Grained Alumina Substrates: I, the Manufacturing Process," *Am. Ceram. Soc. Bull.*, **53** [5] 418 (1974).

3.5 Slip Characterization

32. Micromeritics SediGraph 5100, Micromeritics Instrument Corporation, Norcross, GA.

CHAPTER 4 - THE TAPE CASTING PROCES

4.1 Equipment

1. D. J. Shanefield and R. E. Mistler, "Filter for Ceramic Slips," *Am. Ceram. Soc. Bull.*, **55** [2] 213 (1976).

2. R. B. Runk and M. J. Andrejco, "A Precision Tape Casting Machine for Fabricating Thin, Organically Suspended Ceramic Tapes," *Am. Ceram. Soc. Bull.*, **54** [2] 199–200 (1975).

3. Hirano Tecseed Co. Ltd., Technical Report (II), *Comma Coater and Lip Coater*, Tokyo, Japan, 1995.

4. H. Miyamura, "Recent Trends in Functional Coatings in Japan," Paper, Film and Foil CONVERTECH PACIFIC, pp. 14–18, July, 1997.

5. S. Spauszus and H.-U. Rauscher, "Shaping Principles and Manufacturing Limits of the Production of Thin Ceramic Tapes" (in Ger.), *Keramische Zeitschrift*, **44** [1 23-27 (1992).

6. R. E. Mistler and S. L. Masia, "Coated Paper for Use as a Casting Surface in Tape Processing"; pp. 157–66 in Ceramic Transactions, Vol. 70, *Ceramic Manufacturing Practices and Technologies*. Edited by B. Hiremath, T. Gupta, and K.M. Nair. The American Ceramic Society, Westerville, OH, 1996.

7. C. Lutz and A. Roosen, "Wetting Behavior of Tape Casting Slurries on Tape Carriers;" pp. 163–70 in Ceramic Transactions, Vol. 83, *Ceramic Processing Science.* Edited by G. L. Messing, F. F. Lange, and S. Hirano. The American Ceramic Society, Westerville, OH, 1998.

8. DreiTek, Inc., 1277 Linda Vista Drive, San Marcos, CA, Sales Literature DRE-DS04.

9. R. E. Mistler, D. J. Shanefield, and R. B. Runk, "Tape Casting of Ceramics"; pp. 428–33 in *Ceramic Processing Before Firing*. Edited by G. Y. Onoda, Jr. and L. L. Hench. John Wiley & Sons, New York, 1978.

10. ANSI/NFPA 86, Standard for Ovens and Furnaces 1995 Edition, National Fire Protection Association, 1 Batterymarch Park, Quincy, MA, August 1995.

11. T. Chartier et al., "Tape Casting Using UV Curable Binders," *J. of the Eur. Ceram. Soc.*, **17**, 765–71 (1997).

12. H. D. Lee et al., "Photopolymerizable Binders for Ceramics," *J. Mat. Sci. Letters*, **5**, 81–83 (1986).

13. D. J. Smith, R. E. Newnham, and S. Yoshikawa, "UV Curable Systems for Ceramic Tape Casting," IEEE 7th. Int. Symposium on Applications of Ferroelectrics, Urbana-Champaign, IL, 1990.

14. M. L. Griffith and J. W. Halloran, "Ultraviolet Curable Ceramic Suspensions for Stereolithography of Ceramics," *Manufacturing Sci. and Eng.*, **68** [2] 529–34 (1994).

4.2 Physics and Procedures for Fluid Metering

15. J. L. Park, Jr., "Manufacture of Ceramics," U.S. Patent 2,966,719, Jan. 3, 1961.

16. *CRC Handbook of Chemistry and Physics*; p. F-124. Edited by R.C. Weast. CRC Press, Inc., Cleveland, OH, 1977.

17. Y. T. Chou, Y. T. Ko, and M. F. Yan, "Fluid Flow Model for Ceramic Tape Casting," *J. Am. Ceram. Soc.*, **70** [10] C-280–C-282 (1987).

18. W. C. Moffatt; personal communication.

19. J. S. Reed, *Introduction to the Principles of Ceramic Processing*; pp. 227–52. John Wiley & Sons, New York, 1988.

20. G. A. Somorjai, *Principles of Surface Chemistry*; pp. 72–76. Prentice-Hall, Englewood Cliffs, NJ, 1972.

21. J. L. Moilliet and B. Collie, *Surface Activity*; pp. 97–102. D. Van Nostrand, New York, 1951.

22. J. S. Reed, *Introduction to the Principles of Ceramic Processing*; pp. 22–23. John Wiley & Sons, New York, 1988.

23. W. D. Kingery, H. K. Bowen, and D. R. Uhlmann, *Introduction to Ceramics*; p. 210. John Wiley & Sons, New York, 1976.

24. *CRC Handbook of Chemistry and Physics*; pp. F-45–F-47. Edited by R.C. Weast. CRC Press, Inc., Cleveland, OH, 1977.

25. John Fish; personal communication.

26. R. E. Mistler, D. J. Shanefield, and R.B. Runk, "Tape Casting of Ceramics"; p. 427 in *Ceramic Processing Before Firing*. Edited by G. Y. Onoda, Jr. and L. L. Hench. John Wiley & Sons, New York, 1978.

27. J. A. Lewis, K. A. Blackman, and A. L. Ogden, "Rheological Property and Stress Development During Drying of Tape-Cast Ceramic Layers," *J. Am. Ceram. Soc.*, **79** [12] 3226 (1996).

28. J. C. Williams, "Doctor-Blade Process"; p. 192 in Treatise on Materials Science and Technology, Vol. 19, *Ceramic Fabrication Processes*. Edited by F. F. Y. Wang. Academic Press, New York, 1979.

29. T. A. Ring, "A Model of Tapecasting Bingham Plastic and Newtonian Fluids"; p. 571 in Advances in Ceramics, Vol. 26, *Ceramic Substrates and Packages for Electronic Applications*. Edited by M. F. Yan et al. The American Ceramic Society, Westerville, OH, 1989.

4.3 Drying

30. J. E. Schuetz, I. A Khoury, and R. A. DiChiara, "Water-Based Binder for Tape Casting," *Ceramic Industry*, **129** [5] 42–44 (1987).

31. Publication No. 2008084B, Butvar® Polyvinyl Butyral Resin, Monsanto Company, St. Louis, MO, 1994.

CHAPTER 5—FURTHER TAPE PROCESSING

5.1 Tape Characterization and Analysis

1. E. R. Twiname and R. E. Mistler, "Comparison of Aqueous and Non-Aqueous Tape Cast Nickel Electrodes for Use in Molten Carbonate Fuel Cells"; presented at the 96th Annual Meeting of the American Ceramic Society, Indianapolis, IN, April 25, 1994.

2. R. E. Mistler and D. J. Shanefield, "The Characterization of Unfired Tape Cast Ceramics"; presented at the 1979 Electronics Division Fall Meeting, The American Ceramic Society, Williamsburg, VA, September 1979.

3. S. Forte, J. R. Morris, Jr., and W. R. Cannon, "Strength of Tape Casting Tapes," *Am. Ceram. Soc. Bull.*, **64** [5] 724–25 (1985).

4. A. Karas, T. Kumagai, and W. R. Cannon, "Casting Behavior and Tensile Strength of Cast $BaTiO_3$ Tape," *Advanced Ceramic Materials*, **3** [4] 374–77 (1988).

5.2 Shaping

5. J. A. Miller, "Automatic Handling of Green Ceramic Tapes," U.S. Patent 5,051,219, Sept. 24, 1991.

6. J. L. Park, Jr., "Manufacture of Ceramics," U.S. Patent 2,966,719, Jan. 3, 1961.

7. H. D. Kaiser, F. J. Pakulski, and A. F. Schmeckenbecher, "A Fabrication Technique for Multilayer Ceramic Modules," pp. 35–40, *Solid State Technology*, May 1972.

8. CS-5000 Cutter, Pacific Trinetics Corporation, Carlsbad, CA.

9. Model NCG-3000, New Create Corporation, Osaka, Japan.

10. Model SK-5700 Micro Hole Punching System for Ceramic Tape, Other Thin Materials, Schneider & Marquard, Inc., Newton, NJ.

11. Materials Development Corporation, Medford, MA.

12. Lindberg Heat Treating Company, Rochester, NY.

13. J. R. Piazza and T. G. Steele, "Positional Deviations of Preformed Holes in Substrates," *Am. Ceram. Soc. Bull.*, **51** [6] 516–18 (1972).

14. Model LVS-3012 Laser Via System, Pacific Trinetics Corporation, Carlsbad, CA.

5.3 Calendering and Lamination

15. W. J. Gyurk, "Methods for Manufacturing Monolithic Ceramic Bodies," U.S. Patent 3,192,086, Jun. 29, 1965.

16. G. Zablotny, "Improving Yields in Cofired Ceramic Packages: An Examination of Process and Equipment," *Hybrid Circuit Technology*, **9** [2] 33–35 (1992).

17. Isostatic Laminating Presses, ABB Autoclave Systems, Inc., Columbus, OH.

18. M. Piwonski and A. Roosen, "Lamination of Ceramic Green Tapes"; pp. 424–27 in *Euro-Ceramics V*. Edited by D. Brotzmeyer, M. Boussage, T. Chartier, G. Fantozzi, G. Lozes, and A. Rousset. Trans Tech Publications, Switzerland, 1997.

CHAPTER 6—APPLICATIONS OF TAPE TECHNOLOGY

6.2 Multilayered Ceramic Packages

1. W. H. Liederbach and H. Stetson, "Laminates, New Approach to Ceramic Metal Manufacture: Part II Hermetic Package for Quartz Crystal Oscillators," *Am. Ceram. Soc. Bull.*, **40** [9] 584 (1961).

2. T. Dixon, "Multilayer Ceramics: The Key to High Density Interconnections?" pp. 76–82, *Electronic Packaging and Production*, February 1983.

3. H. D. Kaiser, F. J. Pakulski, and A. F. Schmeckenbecher, "A Fabrication Technique for Multilayer Ceramic Modules," pp. 35–40, *Solid State Technology*, May 1972.

4. J. H. Jean, T. K. Gupta, and W. D. Straub, "Low Dielectric Inorganic Composition for Multilayer Ceramic Packages," U.S. Patent 5,071,793, Dec. 10, 1991.

6.3 Multilayered Ceramic Capacitors

5. G. N. Howatt, R. G. Breckenridge, and J. M. Brownlow, "Fabrication of Thin Ceramic Sheets for Capacitors," *J. Am. Ceram. Soc.*, **30** [8] 237–42 (1947).

6. H. W. Stetson, "Multilayer Ceramic Technology"; pp. 307–22 in Ceramics and Civilization, Vol. III, *High-Technology Ceramics—Past, Present, and Future*. Edited by W. D. Kingery and E. Lense. The American Ceramic Society, Westerville, OH, 1986.

7. W. J. Gyurk, "Methods for Manufacturing Multilayered Monolithic Ceramic Bodies," U.S. Patent 3,192,086, June 29, 1965.

8. G. H. Maher, "Multilayer Ceramic Capacitors for the Next Century"; pp. 91–105 in Ceramic Transactions, Vol. 70, *Ceramic Manufacturing Practices and Technologies*. Edited by B. Hiremath, T. Gupta, and K. M. Nair. The American Ceramic Society, Westerville, OH, 1996.

9. D. Masuda, New Create Corporation, Osaka, Japan; personal communication.

6.5 Fuel Cells for Power Generation

10. R. E. Mistler, "High Strength Alumina Substrates Produced by a Multiple-Layer Casting Technique," *Am. Ceram. Soc. Bull.*, **52** [11] 850 (1973).

11. J. Niikura, K. Hatoh, N. Taniguchi, T. Gamo, and T. Iwaki, "Fabrication and Properties of Combined Electrode/Electrolyte Tape for Molten Carbonate Fuel Cells," *J. Applied Electrochemistry*, **20**, 606–10 (1990).

6.6 Functionally Gradient Materials

12. D. B. Sabljic and D. S. Wilkenson, "Fabrication of Ni–NiO Composites by Tape Casting," pp. 85-87, *Industrial Heating*, September, 1996.

13. L. C. Sengupta, E. Ngo, M. E. O'Day, S. Stowell, R. Laneto, S. Sengupta et al., "Electronically Graded Multilayer Ferroelectric Composites," U.S. Patent 5,693,429, December 2, 1997.

6.7 Separators for Batteries

14. K. A. Blakely, "The Critical Role of Raw Material Suppliers in Market Development" pp. 21–27 in Ceramic Transactions, Vol. 62, *Science, Technology and Commercialization of Powder Synthesis and Shape Forming Processes*. Edited by J. J. Kingsley, C. H. Schilling, and J. H. Adair. The American Ceramic Society, Westerville, OH, 1996.

6.8 Structure-controlled Materials

15. F. V. DiMarcello, P. L. Key, and J. C. Williams, "Preferred Orientation in Al_2O_3 Substrates," *J. Am. Ceram. Soc.*, **55** [10] 509 (1972).

16. "Japan's Synergy Produces Results," *High Tech Ceramics News*, Vol. 7, No. 6, October 1995.

17. S. H. Hong and G. L. Messing, "Development of Textured Mullite by Templated Grain Growth," *J. Am. Ceram. Soc.*, **82** [4] 867–72 (1999).

18. J. A. Horn, S. C. Zhang, U. Selvaraj, G. L. Messing, and S. Trolier-McKinstry, "Templated Grain Growth of Bismuth Titanate," *J. Am. Ceram. Soc.*, **82** [4] 921–26 (1999).

6.9 Rapid Three-Dimensional Prototyping

19. "Lone Peak Engineering Wins Ceramic LOM Process Award," Press Release, Lone Peak Engineering, Inc., September 15, 1992.

20. "Ceramic Tapes in Rapid Prototyping," *High Tech Ceramics News*, Vol 7, No. 9, January 1996.

21. "Ceramic Bones Made to Order," *Industry Week*, Vol. 246, p. 67 (January 20, 1997).

6.12 Thin Sheet Intermetallic Compounds

22. R. E. Mistler, V. K. Sikka, C. R. Scorey, J. E. McKernan, and M. R. Hajaligol, "Tape Casting as a Fabrication Process for Iron Aluminide (FeAl) Thin Sheets"; pp. 258–65 in Materials Science and Engineering, Vol. A258, *Iron Aluminides: Alloy Design, Processing, Properties and Applications*. Edited by S. C. Deevi, D. G. Morris, J. H. Schneibel, and V. K. Sikka. Elsevier, New York, 1998.

CHAPTER 7—WATER-BASED (AQUEOUS) PROCESSING

Introduction

1. G. N. Howatt, "Method of Producing High Dielectric High Insulation Ceramic Plates," U.S. Patent 2,582,993, Jan. 22, 1952.

2. J. L. Park, Jr., "Manufacture of Ceramics," U.S. Patent 2,966,719, Jan. 3, 1961.

3. F. Doreau, G. Tari, C. Pagnoux, T. Chartier, and M. F. Ferreira, "Processing of Aqueous Tape-Casting of Alumina with Acrylic Emulsion Binders," *J. Eur. Ceram. Soc.*, **17**, 311–21 (1998).

4. J. S. Reed, *Introduction to the Principles of Ceramic Processing*; pp.132–51. John Wiley & Sons, New York, 1988.

5. J. Cesarano III and I. A. Aksay, "Stability of Aqueous α-Al_2O_3 Suspensions with Poly(methacrylic acid) Polyelectrolyte," *J. Am. Ceram. Soc.*, **71** [4] 250–55 (1988).

6. Beyong-Hwan Ryu, M. Takahashi, and S. Suzuki, "Rheological Characteristics of Aqueous Alumina Slurry for Tape Casting," *J. Ceram. Soc. of Japan, International Ed.*, **101** [6] 626–31 (1993).

7. N. Ushifusa and M. Cima, "Aqueous Processing of Mullite Containing Green Sheets," *J. Am. Ceram. Soc.*, **74** [10] 2443–47 (1991).

8. C. Pagnoux, T. Chartier, M. de F. Granja, F. Doreau, J. M. Ferreira, and J. F. Baumard, "Aqueous Suspensions for Tape-Casting Based on Acrylic Binders," *J. Eur. Ceram. Soc.*, **18**, 241–47 (1998).

7.1 Water-Soluble Binders

9. A. Kristoffersson, E. Roncari, and C. Galassi, "Comparison of Different Binders for Water-Based Tape Casting of Alumina," *J. Eur. Ceram. Soc.*, **18**, 2123–31 (1998).

10. J. E. Schuetz, I. A. Khoury, and R. A. DiChiara, "Water-Based Binder for Tape Casting," *Ceramic Industry*, **129** [5] 42–44 (1987).

11. E. A. Groat and T. J. Mroz, "Aqueous Processing of AlN Powders," *Ceramic Industry*, **140** [3] 34–38 (1993).

12. Technical Information Form # 192-1064-288, The Dow Chemical Co., Midland, MI.

13. T. Chartier and A. Bruneau, "Aqueous Tape Casting of Alumina Substrates," *J. Eur. Ceram. Soc.*, **12**, 243–47 (1993).

14. K. E. Burnfield and B. C. Peterson, "Cellulose Ethers in Tape Casting Formulation"; pp. 191–96 in Ceramic Transactions, Vol. 26, *Forming Science and Technology for Ceramics*. Edited by M. J. Cima. The American Ceramic Society, Westerville, OH, 1992.

15. P. Blinzer, C. Lutz, A. Roosen, and P. Greil, "Tape Casting of Hydroxyapatite for Medical Applications"; presented at the 101st Annual Meeting of the American Ceramic Society, Indianapolis, IN, April 27, 1999. Poster Number S-AP-001-99.

16. D. J. Shanefield, *Organic Additives and Ceramic Processing*; p. 225. Kluwer Academic Publishers, Boston, 1995.

7.2 Water Emulsion Binders

17. N. R. Gurak, P. L. Josty, and R. J. Thompson, "Properties and Uses of Synthetic Emulsion Polymers as Binders in Advanced Ceramic Processing," *Am. Ceram. Soc. Bull.*, **66** [10] 1495–97 (1987).

18. Dispersion BR Resin Technical Bulletin, Publication No. 6019-D, Solutia, Inc., St. Louis, MO.

19. Butvar® Aqueous Dispersion FP, MSDS Number S00010679, Solutia, Inc., St. Louis, MO.

20. K. Kita, J. Fukuda, H. Ohmura, and T. Sakai, "Green Ceramic Tapes and Method of Producing Them," U.S. Patent 4,353,958, Oct.12, 1982.

21. X. Tang, S. A. Ibbitson, and A. T. Donato, "Acrylic Emulsion Binder for Tape Casting"; pp. 157–64 in Ceramic Transactions, Vol. 62, *Science, Technology, and Commercialization of Powder Synthesis and Shape Forming Processes*. Edited by J. J. Kingsley, C. H. Schilling, and J. H. Adair. The American Ceramic Society, Westerville, OH, 1996.

22. X. Tang; personal communication

23. P. Nahass, W. E. Rhine, R. L. Pober, and W. L. Robbins, "A Comparison of Aqueous and Non-Aqueous Slurries for Tape-Casting, and Dimensional Stability in Green Tapes"; pp. 355–64 in Ceramic Transactions, Vol. 15, *Materials and Processes for Microelectronic Systems*. Edited by K. M. Nair, R. Pohanka, and R. C. Buchanan. The American Ceramic Society, Westerville, OH, 1990.

7.3 Shared Concerns

24. G. A. Somorjai, *Principles of Surface Chemistry*; p. 72. Prentice-Hall, Inc., Englewood Cliffs, NJ, 1972.

Index

U

V

W

X

Y

Z